# A HANDBOOK OF

## Annuals and Bedding Plants

# A HANDBOOK OF

# Annuals and Bedding Plants

## GRAHAM RICE

CROOM HELM
London
TIMBER PRESS
Portland, Oregon

© 1986 Graham Rice
Line drawings by Helen Senior
Croom Helm Ltd, Provident House, Burrell Row, Beckenham,
Kent BR3 1AT

**British Library Cataloguing in Publication Data**

Rice, Graham
  A handbook of annuals and bedding plants.
  1. Annuals (Plants)  2. Flower gardening  3. Bedding plants—
Great Britain
  I. Title
  635.9'312  SB422

  ISBN 0–7099–2277–9


First published in the USA in 1986 by
Timber Press
9999 SW Wilshire
Portland, Oregon 97225
All rights reserved

ISBN 0–88192–029–0


Typeset by Oxford Publishing Services, Oxford
Printed and bound in Great Britain by Butler and Tanner Ltd, Frome

# Contents

# Colour Plates

1. *Antirrhinum* 'Cinderella' is one of a new range of antirrhinums which are almost as good as the F1 hybrids but at a much more reasonable price. (Photograph: Suttons Seeds)
2. *Rudbeckia* 'Goldilocks' was bred by Hursts in Britain and won a Fleuroselect Bronze Medal. (Photograph: Fleuroselect)
3. An attractive combination of flower and foliage. *Amaranthus* 'Red Fox', *Pyrethrum ptarmicaeflorum*, *Heliotropum* 'Marine' with *Iberis* 'Giant Hyacinth Flowered White' in the background. (Photograph: Dobies Seeds)
4. The mountain pansy (*Viola lutea*), growing here in Aberdeenshire, brought a more tufted habit to garden pansies. (Photograph: Phil Lusby)
5. 'Majestic Giant' pansies are widely grown for both summer and spring flowering. (Photograph: Dobies Seeds)
6. The wild heartsease (*Viola tricolor*) was one of the most important progenitors of the garden pansy. Here it is growing through a purple-leaved maple. (Photograph: Author)
7. Vegetables and flowers making a striking combination: Swiss chard with *Verbena* 'Blaze'. (Photograph: Author)
8. Corn marigold (*Chrysanthemum segetum*) growing on the Black Isle, Wester Ross, Scotland. It also makes a big splash in the annual border. (Photograph: Phil Lusby)

9. Red, white and blue in a garden trough. Geranium 'Scarlet Diamond', *Cineraria* 'Silverdust' and *Lobelia* 'Sapphire'. (Photograph: Rod Sloane, *Practical Gardening*)
10. An unusual combination of yellows: marigold 'Inca Orange' with *Chrysanthemum* 'Gold Plate'. (Photograph: Rod Sloane, *Practical Gardening*)
11. Single colours can be very effective in hanging baskets. *Calceolaria* 'Sunshine' flowers all summer. (Photograph: Rod Sloane, *Practical Gardening*)
12. A mixture of colours from just one plant, in this case *Begonia* 'Non Stop', gives consistency of habit with a variety of flower colour. (Photograph: Rod Sloane, *Practical Gardening*)
13. Colours that almost match with a little contrast added. 'Orange Bedder' wallflowers, orange 'Halcro' tulips with a white tulip. (Photograph: Author)
14. The tiny *Ionopsidium acaule* is an unusual spring bedder and needs a dwarf tulip like this *Tulipa greigii* 'Cape Cod' to go with it. (Photograph: Author)
15. Geranium 'Scarlet Diamond' is the first of the new 'multiflora' geraniums and one of just three Fleuroselect Silver Medal winners. (Photograph: Fleuroselect)
16. A modern cornflower variety, in this case 'Polka Dot'. (Photograph: Dobies Seeds)
17. Cornflower (*Centaurea cyanus*) growing wild in a barley field on the Black Isle, Wester Ross, Scotland. (Photograph: Phil Lusby)
18. One of the most beautiful of all plants is *Ipomoea* 'Heavenly Blue', here in front of *Lavatera* 'Silver Cup'. (Photograph: Author)
19. Three silver foliage plants in different styles. *Cineraria* 'Silverdust', *Cineraria* 'Cirrus' and *Centaurea gymnocarpa*. (Photograph: *Garden Answers*)
20. The huge-fingered foliage of *Ricinus* 'Impala' makes a good contrast with *Nicotiana affinis*, the best of the scented tobacco plants. (Photograph: Author)

# *Figures*

# *Introduction*

Anyone who fights against change is going to have an unhappy life. Life *is* change and the sooner we learn to like it the happier we will be. Combine this with one of my other favourite maxims — that nothing is ever quite perfect, never will be and we will be constantly dissatisfied if we think otherwise — and you have a good reason for growing annuals and bedding plants. They give a display for just one summer or one spring so each season and each year you can give your garden a dramatic new look, yet retain the elements which particularly please you by simply planting them again. Those plants and schemes which are less satisfactory can be dispensed with and others put in their place — only some of which will work. Add the unpredictable nature of the climate and you will find that your garden is in a constant state of change, of improvement, and causing both frustration and great satisfaction. Never a dull moment.

This is a book for the home gardener so I have only explained techniques which are practicable in the domestic situation. Likewise, the plants which I have described or recommended are almost entirely those which are available to the home gardener in garden centres or in mail order catalogues. Varieties only available to the professional grower have been almost entirely excluded. Only when I know or suspect that they will soon be introduced for the amateur have I allowed them to make an appearance. Varieties which I suspect may soon disappear have largely been excluded. However, I am sure that professional growers will find much of interest; especially, I hope, the inspiration to grow more of the less common plants which home gardeners already grow.

Within the parameters of this book I have concentrated on plants raised from seed, although I have not hesitated to give you ideas where growing from cuttings can be used to make the most of seed-raised plants. I have therefore not discussed in depth such plants as fuchsias, cuttings-raised geraniums, cannas and iresines, and the wonderful *Helichrysum petiolatum* and its varieties — indispensable plants for containers. However, I am sure that there is plenty here to inspire you to try something new and try it with success.

I have suggested a number of plant associations throughout the book and this can be risky. Tastes vary dramatically so I have ensured that the suggestions do not represent my taste to the exclusion of others. But you will soon become aware of where my preferences lie. There will be ideas for planting with which you disagree but some of these will appeal to others and overall everyone should find schemes they like. There are really no rules; if you like a combination, then it is a good one.

Plant names can be confusing for everyone. Academic botanists, for perfectly good scientific reasons, may change the name of a familiar plant. Should it be called *Senecio maritima* 'Silverdust', *Cineraria maritima* 'Silverdust' or *Senecio cineraria* 'Silverdust'? Rather than follow strict botanical correctness, which would provoke irritation in most readers, I have tried to use the names which are most often used in catalogues. This makes it simpler for everyone. Then there is the problem of seed companies changing the names to suit themselves. The geranium 'Bronco', probably the best traditional-style red geranium from seed, is also often seen as 'Solo'. This is simply confusing and unnecessary. There is also the famous case of *Begonia* 'Lucia' which has been given three other quite different names by different seed companies because they have slightly altered the colour balance of the mixture. How much simpler if they had merely called them 'Dobies Lucia' or 'Suttons Lucia' instead of 'New Generation' or 'Devon Gems'; at least we would then have a clearer idea of what we are buying.

But this is just a small complaint and should not detract from the fun of growing these plants and in particular the fun of trying out something new each year in an effort to create perfect displays — but always being delighted with just a little less than absolute perfection.

# *Acknowledgements*

Many people have fed my interest in seed-raised plants and I give them my warm thanks. First Brian Halliwell, an Assistant Curator at Kew, whose brave and inspired approach to bedding is always refreshing; and also Paul Hansord, once of Unwins Seeds, whose enthusiasm is so infectious. And then later, in no particular order, Tom Sharples and the inimitable 'Rad' Radaway of Dobies, Keith Sangster and David Tostevin of Thompson and Morgan, David Kerley and Andrew Bradley of Unwins, David Haswell of Asmer, Mike Hough of Floranova and Tony Byers of Suttons.

My thanks are due to the following for permission to reproduce photographs or use them as references for artwork: *Practical Gardening,* Phil Lusby, Dobies Seeds, Unwins Seeds, Thompson and Morgan Seeds, Suttons Seeds, *Garden Answers,* Fisons Horticulture, Geoff Hamilton, the Harry Smith Collection, P.R. Chapman, Bernard Alfieri, Fleuroselect. The excellent line drawings are by Helen Senior.

My mum has been an enormous help in suggesting many small but significant improvements to the text as well as pointing out the occasional howler. And thanks also to Jo Hemmings, my editor at Croom Helm, who combined both sympathetic tolerance and cheerful chivvying in order to ensure this book appeared on time.

Finally, my thanks are due to Mike Wyatt, my editor at *Practical Gardening*, who has allowed me not only to use material which previously appeared in the magazine and pictures from his files, but who also turned a blind eye on my rather bleary arrival after I had been working on the manuscript late the night before.

# CHAPTER 1
# *Raising the Plants*

## Seed and Cuttings

All annuals and the vast majority of bedding plants are grown from seed. Even plants like geraniums and gazanias, which used to be grown from cuttings, have now developed as seed-raised plants. This has been done for three reasons. First, disease was often perpetuated by constantly taking cuttings and the presence of virus diseases in particular led to deterioration of the resultant plants in the form of poor flowering capacity, generally weak growth and, paradoxically, a reduced number of cuttings actually rooting. The cost of overwintering these tender plants in greenhouses became a more important factor after the oil crisis when the cost of fuelling boilers increased dramatically and finally, with most bedding plants being seed-raised, growing plants from cuttings began to fit less well into the production schemes of large commercial growers. Plant breeders have put a great deal of work into breeding seed-raised types and geraniums, gazanias, chrysanthemums, dahlias, impatiens and coleus have now been developed as high quality seed-raised plants.

Recently, with greatly improved ideas on disease control and greenhouse insulation, there has been a swing back to growing plants through the winter, either from seed as an energy-saving technique, or from cuttings in order to perpetuate forms which cannot be raised from seed or which have been selected from mixtures. This last reason is one of the most useful advantages of growing from cuttings for the home gardener.

Many plants are only available to the home gardener in mixtures — commercial growers and parks departments are luckier — so home gardeners, if growing plants which are inherently perennial, must go about things in a different way. A mixture of fibrous rooted begonias may contain quite a wide range of flower and foliage colours, one of which may be especially appealing or suited to a particular spot in the garden. One or two plants from the mixture can then be selected out, dug up and potted at the end of the season, overwintered and used as sources of cuttings from which to grow more plants the following spring.

## Buying Seed

Mail order catalogues give you the best selection of varieties — especially if you send for them all and pick your favourites from each range. Unfortunately, realistic price comparison is not possible as in Britain only one company actually states how many seeds each packet contains.

*Figure 1.1 Some seeds come packed in metal foil. Not only do they stay viable for longer, but when they germinate (left) they do so far more vigorously than seed stored in paper packets (right)*

F1 hybrid seeds are usually marked as such and are more expensive. An F1 hybrid is the result of a cross between two specially created parents which are quite uniform but may themselves not be good garden plants. When the cross is made, often by hand, the combination of genetic material in the new plant is very predictable and will yield a vigorous, floriferous plant which is early flowering and very uniform. F1 hybrids are expensive because the seed stocks of the two parent plants must be maintained to a very high standard, and of course the labour costs of the pollination and harvesting may also be high.

## Sowing Seed Outside

Both spring bedding and biennials are sown in one place, then transplanted to another for flowering. The only difference is that some spring bedding plants are actually perennials and although they will flower from year to year, are treated as more temporary plants as their flowering capacity falls after the first season. Wallflowers, polyanthus, pansies, all seed-raised plants, fit into this category. Except for some F1 hybrid varieties, the seed of which is very expensive, all are sown in the open ground in summer and transplanted in the autumn for flowering in spring.

**Spring Bedding and Biennials**

The site for the nursery bed is often forced upon the gardener by the size and nature of the space available when sowing time comes around. It is often most convenient to fit the seed bed into the vegetable area where it can be built into the normal rotation. This is especially useful for wallflowers which suffer from the same dreaded clubroot disease as cabbages and other members of the brassica family — so put them in the same part of the allotment. The site should be open, preferably sunny, but if some shade lovers like primroses and violas are likely to be grown, a partly shaded spot is often the most useful. In practice, anywhere which is not actually overhung by trees will prove satisfactory. If it can be sheltered from strong winds this will help you develop sturdy plants from a slightly later sowing than would otherwise be the case.

Again the soil in the vegetable area usually proves suitable — fertile, but not over-rich. Most importantly it should have good water holding capacity so that in the summer months, when the seedlings are small and at their most vulnerable to drought, there will be no need for constant watering. Both acid and alkaline soils grow good biennials. For a plant to become universally popular, as so many biennials are, it has to be tolerant of a wide range of conditions and as long as the organic content is not too low, any other features can usually be tolerated or adjusted during the sowing routine.

There is argument about the best time to sow spring bedders, especially wallflowers. I have rather changed my mind over the years. Preferences vary between May and July and at first I was inclined to leave sowing until July. This was reinforced when I had a garden in the East Anglian Fens as the larger plants from a May or early June sowing were so caught by the autumn gales that the display in spring was less impressive than it should have been. Gardening now in a more sheltered spot, I can see the virtue of sowing earlier and by putting out larger plants there is definitely a better show in spring. This can be summarised by recommending that you plant out the largest plants that will thrive in your particular garden and time your sowing accordingly. Start with a June sowing and see what happens. But remember that the larger the plant you set out, the more important it is that you dig it up carefully and keep plenty of soil on the roots.

Seed for spring bedding should be sown in a well-prepared seed bed in short rows, one row for each colour or variety.

(1)  Assuming the site to be on reasonably fertile cultivated ground, fork over lightly, adding some moist moss peat if the soil is on the sandy side and some peat or sharp sand if it is especially heavy. You should need no fertiliser at this stage unless the soil is especially impoverished in which case 1oz of superphosphate per sq yd (60g per sq m) should be raked in.

(2)  Tread the area with the heels of your boots and rake level.

*Figure 1.2 When raising a number of different spring bedders, set aside a small area in which to sow short rows of seed*

(3) Write the labels for the plants you are sowing and put them in 6in (15cm) apart.

(4) The length of the rows you sow depends entirely on how many plants of each you need, but for most gardeners a row 1–2yd (0.9–1.8m) long is adequate. Using a cane or board as a guide, mark out a drill about ¼in (6mm) deep behind each label with the point of a cane.

(5) If the weather is very dry at sowing time, make the drill twice as deep and then flood the drill with

*Figure 1.3 When sowing seed of spring bedding plants in dry conditions, water the drills first to make sure the seed does not lie in parched soil*

water from the spout of a watering can and let it soak in before sowing.

(6) Sow the seed thinly, three to six seeds to the inch (2.5cm) is about right, depending on the reliability of the plants involved.

(7) Use the back of the rake to draw soil over the seeds and gently tap the soil over them with the flat of the rake. There is no need to give different plants different treatment at this stage — forget-me-nots, pansies, bellis, wallflower can all be treated the same way. Only the larger plants like cotton thistle or foxgloves will need more space.

(8) When the seedlings have produced their first true leaves, as distinct from their first rather simple, uncharacteristic pair of seed leaves, they can be thinned out a little. Remove only enough to leave seedlings 1–1½in (2.5–4cm) apart. Water the rows

*Figure 1.4 When
thinning seedlings
or removing weeds
from the row, put
your fingers around
the seedling you
want to retain to
keep it in place*

first, then, placing two fingers around each seedling you want to retain to keep it in place, remove the others.

(9) When the seedlings have put on some more growth they should be transplanted to give them space to develop properly. Water the rows, lift the plants carefully with a handfork, lay them in seed trays and cover them with polythene to stop them drying out.

(10) Plant them in a nursery bed prepared in a similar way to the seed bed, but if the fertility is low rake in 2oz of Growmore per sq yd (60g per sq m) before planting. The spacing depends on the size of plants you are trying to produce, but 6–9in (15–23cm) for honesty and wallflowers and 4–6in (10–15cm) for smaller plants like bellis and pansies is about right. Keep as much soil on the roots as possible and water them in well straight after planting. If the weather is hot, some shading for a few days will be very useful.

(11) Water as necessary for the rest of the summer and keep an eye out for pests and diseases, although they are rarely troublesome.

(12) In the autumn, after the summer plants have been removed from the garden, the spring plants can be transplanted to their final positions. Again, water the beds well, lift carefully and take as much soil

on the roots as possible. Just because you see wallflowers in markets with their bare roots in buckets of water does not mean that it is the best way to treat them. Plant them out a little further apart, by about one-third, than they were in the rows.

A large number of the plants described in this book are hardy annuals, plants which can be sown outside in spring for flowering the same summer. Unlike biennials, they are usually sown where they are to flower and simply thinned out to give space for them to grow.

**Summer Flowers**

Most hardy annuals like an open, more or less sunny site. This is related directly to their habitats in the wild — you will not find many annual plants growing in woodland. There are some which really do demand sunshine all day long and these are indicated in the alphabetical section of the book. Most are happy with sunshine for more than half the day and a few will thrive on rather less. Shade from a fence or house wall is far more acceptable to most plants than shade from trees; the latter is likely to be from overhead and so lets through much less light.

There is a great and popular rumour that annuals like nothing better than a parched, impoverished soil. This is rubbish. What you will get if you sow in such a poor soil is early flowering, which will be relatively dramatic because the plants will be small. Then they will set seed and die and that will be the end of your summer flowers. This roughly reflects the way in which many annuals grow in their wild homes where seed lies dormant in the soil for a large part of the year and then suddenly grows and flowers after rain, going through the whole cycle quickly to make the most of the available water.

In cultivation we need to manipulate things a little to get the best from our plants. Most of the entries in the alphabetical section include a few words on the best soil but in general the soil should be such that the plants can get as much plant food as they need in the right balance, and should provide the capacity to hold as much water as they need to keep them growing all summer.

The plant food balance is important, the one thing you do not need is too much nitrogen — this is the element which encourages leafy, vegetative growth, usually at the expense of flowers. Of course, a soil with no nitrogen, if such a thing were possible, would result in almost no growth at all. But potash is the element that is most important for flower production. Fortunately, it so happens that most garden compost is relatively low in nitrogen and high in potash, as is spent mushroom compost. Farmyard manure and poultry manure are the ones to treat with caution. What I do, trying to keep things simple, is to dig in garden compost or mushroom compost, and then rake in 2oz per sq yd (60g per sq m) of Growmore before sowing. Growmore is a balanced fertiliser which contains equal amounts of all three major nutrients — nitrogen, phosphate and potash. Sulphate of potash at the same rate would be better and you could also argue that superphosphate is useful in the early stages as it encourages a good root system. But in order to keep the shed free of half empty sacks of fertiliser which sit there for years, I reduce everything to absolute basics and get one or two enormous sacks of Growmore to use for just about everything. My simple system gives the plants a good water-retentive soil to help them cope with those summer dry spells that take us by surprise every few years, enough nitrogen to get them going well without creating big leafy bushes and a sufficient balance of other nutrients to encourage enough root growth and so make the most of the water reserves, resulting in plenty of flowers — which after all is the point of the exercise.

In southern parts of Britain, most hardy annuals can go in during the second half of March. Leave it a little later as you go further north but always look at the weather and the soil conditions rather than the calendar. If it is very wet or very frosty then forget it, your plants will flower no later if you wait. Sow too early and you may end up sowing again anyway. Some annuals, and these are mentioned later, are a little less hardy than others and can usefully be left until April as a matter of course.

There are also a number of plants whose natural life cycle includes germinating in late summer and overwintering as ground-hugging rosettes before flowering the following year — not unlike a biennial. They will flower earlier than spring-sown plants but usually after the biennials. They are best sown in late August or early September in most areas although some success has been had by sowing as early as late July in the South of Britain.

When sown in summer, some annuals are likely to flower well into the autumn. Suttons Seeds have carried out trials over a number of years on which annuals are best treated in this way. They have come up with this list of varieties, most of which, if sown in the third week of July at their trial ground in South Devon, will flower from the end of August and into October.

| | |
|---|---|
| *Centaurea* 'Polka Dot' | *Godetia* 'Dwarf Selected' |
| *Clarkia pulchella* | *Iberis* 'Fairy Mixed' |
| *Delphinium* 'Dwarf Rocket' | *Linaria* 'Fairy Bouquet' |
| *Dimorphotheca* 'Glistening White' | *Nemophila insignis* |
| *Eschscholtzia* 'Ballerina' and 'Miniature Primrose' | *Reseda* 'Sweet Scented' |

There are two methods of sowing hardy annuals where they are to flower, but the soil preparation is the same for both.

(1) Clear the ground of any overwintered weeds and, if you have not done so in the previous autumn, dig in your organic matter. You will not need to do this every year, it rather depends on the inherent state of the soil's fertility, but make sure your compost really is well rotted and in that case, rather than leave it in a layer at the bottom of the trench, fork it in more evenly.
(2) Tread well, rake in your fertiliser and you are ready to sow.
(3) If you are sowing a number of varieties together in a bed, mark out the area into patches using a cane or sand on the raked surface.

*Figure 1.5 Mark out areas for sowing hardy annuals by spreading lines of sharp sand on the prepared soil*

9

(4) This is where the methods diverge.

At Kew, seed of annuals is sown broadcast in patches to give an informal, well knitted appearance. This is only advisable if you can recognise the most common weeds at the seedling stage — or at least distinguish them from your plants.

(a) Using the rake, draw soil from halfway across the patch to one side making a low ridge at the edge and do the same in the opposite direction, leaving a ridge on that side too. This will give you a slightly sunken area which should be easy to rake level and finely, with a ridge at each side. The sunken sowing area should be ½–1in (13–25mm) deep. The seed is then scattered evenly over the whole area. This is usually best done by tipping the seed into the palm of the right hand, for right-handed gardeners, slightly closing the hand to give a crease in the skin and tapping the wrist bone with the index finger of the other hand to encourage the seed to flow along the crease and on to the soil. This technique is especially useful for medium- to small-sized seed and gives you the great advantage of being able to see the seed as it falls from your hand. Most seed is too dark to be seen on the soil but by watching it fall from your hand you can get a good idea of how evenly you are sowing.

(b) After sowing, the soil from the ridges at the sides is used to cover the seed. This is best done by using the back of the rake to flick soil from the ridges across the sown area in the first instance so that the newly sown seed is disturbed as little as possible. The remainder of the soil can then be drawn across and levelled carefully.

(c) The whole area is then firmed by gentle tapping with the back of the rake.

(d) It is when the seedlings come through that problems could arise. In practice it will usually be apparent, simply by looking at the numbers present, which are weeds and which are your seedlings. When the seedlings are ½–1in (13–25mm) high, weed seedlings can be removed and any seedlings which are especially crowded also thinned. Depending on the eventual plant spacing, which is to be found on the seed packet and will depend on the variety involved, another thinning or two will be needed. For large plants needing an eventual spacing of 9–15in (13–38cm) it pays to thin out in stages to avoid gaps due to unavoidable losses.

The alternative is to sow in drills. This can be simpler but leads to a more regimented appearance, especially in the early stages. It enables you to spot the weeds easily but you will probably need more seed.

(a) Using the corner of the rake or a cane, draw drills the distance apart specified on the packet. They should be around ½in (13mm) deep. Seed is sown thinly along the drill aiming for about three or four seeds to the inch (25mm); if in doubt, err on the generous side. The seed is then covered and firmed gently, using the back of the rake to draw soil over the drills and tamp it down.

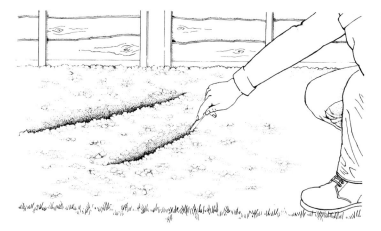

*Figure 1.6 Mark out drills for sowing with the point of a cane*

(b) When the seeds come through weeds can be removed from the areas between the drills. A hoe can be used if the spacing is wide, and weed seedlings removed from within the rows too. The seedlings themselves are thinned out in stages to the spacing recommended on the packet.

Many half-hardy annuals, like marigolds and zinnias, can also be sown outside where they are to flower if left a little later than the hardy types. Even geraniums will flower in their first year from seed, just, if sown in a sheltered spot in late April but it is hardly to be recommended. But by

11

waiting until later in the spring than you would sow the hardy types, you can still have the pleasure of growing these less hardy ones even if you have no greenhouse.

Some are hardier than others. For example, antirrhinums and eccremocarpus are hardier than zinnias and ipomoeas so can be sown slightly earlier, but for most plants the first week in May is about right and if there are any frosts forecast at the end of May when the seedlings are through, a sheet or two of newspaper is enough to protect them. Some, marigolds especially, will even tolerate transplanting and as they are so quick growing you can sow in June and move plants to fill gaps in beds caused by accidents.

## Sowing Seed Inside

Most hardy annuals are not difficult to raise and those half-hardies which are sown outside are fairly straightforward too. But although many people have problems with raising half-hardy annuals in the greenhouse or on the windowsill, this can be a reliable method providing the rules are followed. There is a variety of techniques available and a wide range of equipment.

**Equipment**

Ideally, a greenhouse or conservatory should be the basis for your plans. Any greenhouse is better than none, but it must be said that recent models do offer advantages in

*Figure 1.7 It is surprising how many seedlings and young plants you can fit into a greenhouse*

12

design. Larger panes of glass to let in more light, improved draught-free glazing techniques, narrower glazing bars, potential for full ventilation, a wider range of sizes and permutations of design, easier erection, and full provision for the attachment of accessories. Most of these improvements have been to aluminium greenhouses but with the price of aluminium going up, more thought is going into timber greenhouse design too.

Possibly the most important development as far as structural design is concerned is the increased ease of adding extensions and partitioning off parts of the greenhouse. This and many other developments amongst the accessories relate to the cost of heating.

Fairly high temperatures are needed not only for the germination of bedding plants but also for the early stages of their growth and it costs a lot to keep a greenhouse heated in the late winter and spring so it makes sense to heat the smallest area possible. This is where a partitioned greenhouse is very useful. If one small part of the greenhouse is partitioned off it can be heated to a higher temperature and used for germination and early growth and then the plants transferred later to the body of the house which is kept cooler. Almost the whole of the small partitioned area can be filled with staging and the gardener can stand in the doorway to tend to plants, there is no need for an aisle. If the staging is such that there is a lower shelf below the main level and narrow shelves fixed to the sides, an enormous number of seedlings can be raised. This set up is also ideal for overwintering tender perennials like fuschias and geraniums in just frost-free temperatures.

A further development of this idea is the use of a heated propagator. This can be sited in the small, high temperature area of the greenhouse and used solely to germinate seed. It is at the germination stage that the highest and most constant temperatures are needed and so the propagator can be filled with seed pots and as soon as germination starts in one pot, it can be moved on to the staging and replaced with another new sowing. But a propagator with a thermostat is essential. This will cost more but will keep your seeds at exactly the temperature you set regardless of the weather and other variables. Most importantly it will ensure that on sunny days when a propagator can warm up very quickly, the heating element is not adding to the warmth and keeping the seeds too hot. The other piece of advice is to buy a larger one than you think you need; you will have no trouble filling it and of course it can be used for cuttings as well as reviving ailing plants. Lastly, make sure it has an adjustable thermostat so you can keep it at the temperature you want, not the temperature the manufacturer sets in advance.

When it comes to pots and trays plastic is the rule these days, plus compressed peat. The days of clay pots are not entirely over, but these very useful clays are best kept for plants that really benefit from their porous sides like succulents and alpines. Otherwise it is plastic.

For sowing seeds, pots are preferable to trays. Most home gardeners do not need enormous quantities of any one variety so sowing a tray or even a half trayful is wasteful. And dividing a tray up and sowing two or three varieties in it is asking for confusion. So instead, use 3½in (9cm) pots for most varieties if you only want a couple of dozen plants or 5in (12.5cm) half pots if you need more. If you can find square pots for seed sowing rather than round ones, and they do exist though are difficult to find, you will save space in the greenhouse. When you sow in these pots, there is no need to fill the entire pot with seed compost; the bottom half can be filled with peat, which is cheaper, as the roots of the seedlings will probably not penetrate that far before you transfer them.

For growing on there are many options. The standard container is the seed tray, traditionally in wood, and still

*Figure 1.8 If necessary, many kinds of containers can be pressed into service for pricking out*

available but far more common in various plastics. The one big problem with wood is that when you have used it once it is very difficult to get it as clean and sterile as you need. You can still get tomato boxes from the market or greengrocer but they are unnecessarily deep for most seedlings. So plastic trays and half trays are the norm. The types of plastic of which they are made vary enormously but I would suggest that the flimsy ones be avoided. They are cheaper, to be sure, but unlikely to be of use for more than one season unless you are very careful with them and they are difficult to carry without their breaking. I would suggest instead that you buy the most sturdy you can, even if they do cost more. They will last well and they are easy to clean. And make sure you have full size and half size trays to give you the flexibility you will need if you are growing a variety of plants.

Some plants are best pricked out into pots and 3 or 3½in (7.5 or 9cm) square plastics are the best bet; they take up the least space. There are various cell packs which are also useful. Cell packs are small pots linked together into units which fit into seed trays which act as sturdy holders. These

*Figure 1.9 Propapacks are very useful for avoiding the necessity of pricking out seedlings — you sow the seed direct*

*Figure 1.10 The Propapack is pressed on its upturned base and the plugs of soil with their seedlings are pressed out*

15

give you the advantage of growing each plant separately in its own ball of compost yet they can all be carried about easily. A variation is the Propapack, a block of polystyrene divided into 24 or 40 cells.

One or two plants, like eucalyptus, which especially dislike root disturbance, can be pricked out into pots made of compressed peat and the whole lot planted out so that the roots grow through the peat into the soil. The

*Figure 1.11 When the roots of seedlings in peat pots penetrate the base, they should be potted up*

alternative is to sow in cells in the first instance. Vigorous varieties can go in larger cells, slower growing types in smaller cells or in small peat pots. One or two seeds are sown in each cell, thinned to one and then left to grow on.

A very useful item for the home gardener which in many ways is easier to manage is the Jiffy 7. This is a cylinder of compost surrounded by a thin plastic net to keep it together. When you take one out of the box it is about 1½in (4cm) across and ¼in (6mm) deep. As it soaks up water it expands vertically until it is about as tall as it is wide — there is also a useful dimple in the top. Seed is sown on the top, thinned to one seedling in the case of fine seeds, and the whole thing moved into a pot for growing on before planting. It is important to use them in conjunction with a capillary watering system — matting or sand. The Jiffy 7s are set out almost touching on the moist mat and watered overhead with a fine rose on the can. Overnight, they will swell up and be ready for sowing. They are ideal for geraniums and zinnias, and can also be used for cuttings, which are simply pushed into the top.

Compost is a subject upon which everyone has their own very definite preferences. Generally speaking the peat-based composts are more reliable but even so you can still get a surprise when you open the bag — the quality of some brands is not consistent.

However, there are general observations which can be made. First, always choose a brand that includes sand, grit, perlite or vermiculite as well as peat. These additional constituents are added to improve the drainage of the compost and with the peat element usually being the variable factor, good drainage is especially necessary. Most peat used in proprietary compost is milled to a very fine grade and this too can impede drainage when the peat settles or is firmed too much. The addition of this drainage material also helps the peat to absorb water more easily if it is allowed to get very dry.

You can also mix your own, using fine grade moss peat and perlite or vermiculite plus a special fertiliser. All the materials will be quite sterile when you buy them so it pays to mix them on a clean board to ensure that the compost is not contaminated by weeds or fungi which could damage your seedlings. There have been experiments on the use of used growing bag compost for bedding plants but the unpredictable nutrient levels and the physical condition of the compost ruled it out.

The basic constituents of mix-your-own compost are medium grade moss peat, perlite or vermiculite and Chempack Seed Base or Potting Base. These two base fertilisers contain all the nutrients and trace elements needed to feed healthy seedlings and plants. The peat should be sifted through a ¼in (6mm) sieve to remove any lumps; alternatively fine grade peat can be used. The perlite should be of horticultural grade, as should the vermiculite. It is most important that the peat be moist when mixing takes place and that mixing be thorough. This compost is not only very effective as a growing medium but is far cheaper to make than proprietary composts are to buy.

**Routine**

The sowing routine is straightforward enough, although there are important details which should not be ignored.

(1) Start with impeccably clean pots, bench and hands plus fresh compost.

(2) Write the labels for the packets you want to sow.

(3) Water the compost in good time, a couple of hours before sowing, so that it is uniformly moist but not so wet that a fistful drips when squeezed.

(4) Open the first packet, see how much seed is in the packet and think how many seedlings you need — then decide the size of the pot or tray you need. (Unused seeds can be stored in a tin in a cool, dry place).

(5) If you are using a pot or pan, half fill with loose, moist peat and then fill with your seed compost. Brush off any surplus with the flat of the label, tap

*Figure 1.12 When filling seed trays, make sure the compost is firmed evenly and do not neglect the corners*

the pot on the bench once or twice then use the bottom of another pot or a specially made presser to firm the compost gently, leaving between ¼–½in (6–13mm) at top.

(6) Now sow the seed thinly over the whole surface of the compost. Large seeds can be spaced out or

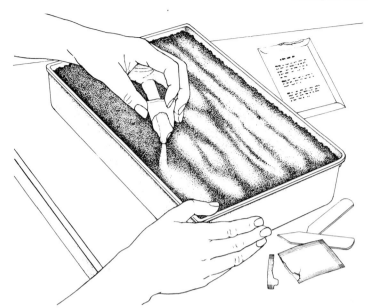

*Figure 1.13 One way of sowing very fine seeds evenly is to mix them with sharp sand*

moved to a more even spacing with the tip of a label after sowing. The sowing of especially fine seed is dealt with under begonias on page 94.

(7) Most seeds should be covered with their own depth of compost but fine seeds should not be covered at all, or by a very small amount of compost. It is essential to use a sieve to cover the seeds, the old

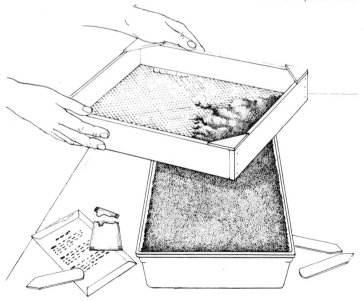

*Figure 1.14 Use a purpose-made sieve, or an old coarse mesh kitchen sieve, to cover the seed evenly*

technique of rubbing compost between the palms gives a very heavy and uneven covering. Special seed sieves are available or you can use a coarse kitchen sieve. For fine seeds simply hold the sieve over the pot and give it a tap.

(8) The seeds should be watered in and this is best done with a watering can equipped with a very fine rose. It is often suggested that pots be stood in water

*Figure 1.15 When watering in seeds and seedlings, use a can fitted with the finest of roses*

then removed and allowed to drain but as the pots drain the compost tends to pull away from the sides of the pot and to sink. This is especially likely with peat-based composts. Always add a copper fungicide such as pbi Cheshunt Compound (copper sulphate and ammonium carbonate) to the water to help prevent damping off.

(9) Put the seed pots in the propagator with the thermostat set at 70°F (21°C) and cover the lid with newspaper. If you intend to keep seedlings growing in the propagator after germination, the pots will need to be covered individually with a square of glass or perspex and brown paper or newspaper.

(10) As soon as germination starts, the covering should be removed or the pots moved to the bench.

(11) As soon as the seedlings are big enough to handle they should be pricked out. Water the pot and then fill the pots or tray with compost. A similar compost to that used for seed sowing should be used.

The procedure of pricking out, moving seedlings from their crowded seed pots to seed trays where they have more space, is often a hazardous one. The seedlings should be moved as soon as they are large enough to handle and should only be gripped by the leaf between thumb and forefinger to minimise damage. Try to damage

*Figure 1.16 Handle seedlings very carefully, making sure not to damage the delicate stems*

roots as little as possible by easing the seedlings out with a dibber and making a good sized hole in the compost into which the seedling is being moved to avoid squashing the

*Figure 1.17 Firm seedlings in gently, but take care not to damage fine roots*

roots. Firm gently and then water well with a drench of pbi Cheshunt Compound (copper sulphate and ammonium carbonate) straight away. The spacing in the tray will depend on the size and vigour of the plants involved; from seven by five in a half tray for lobelia to four by three in a half tray for larger more vigorous plants like bedding dahlias.

*Figure 1.18 If you prick out a lot of bedding plants, a homemade presser is worthwhile*

Recent experiments have shown that many plants are quite happy if the temperature is lowered to 40°F (4°C) immediately after pricking out. Of course growth will be slower but the fuel saving will be substantial. Very careful attention to watering is crucial and regular sprays against botrytis will be useful, especially on slower-growing plants like begonias.

Another interesting outcome of this research, at Lee Valley Experimental Horticulture Station, shows how differences in sowing date are compressed when it comes to maturity. Batches of aster, for example, sown six weeks apart, were only one week apart by planting out time. So do not be afraid to delay sowing.

After pricking out feeding will help build a sturdy plant quickly. A general feed such as Liquid Growmore is perfectly adequate for seedlings at every stage until planting out but a feed with a little more nitrogen can be used in the early stages, especially if sowing has been late and you need to encourage growth to catch up. But do not overdo the nitrogen or the plants will be difficult to harden off well and they will be more susceptible to pest and disease problems.

Commercial growers are discovering that they can save fuel by sowing in the autumn and keeping the plants cool, rather than sowing in spring and growing at much higher temperatures. Improved insulation has helped keep the cost down, improved fungicides are helping to deal with the disease problem and improved varieties which do not get too leggy in the low winter light make the whole business possible. At the moment it is mainly geraniums and *Campanula isophylla* which are treated in this way. It will probably be worth trying the same idea for other plants, especially those like begonias that need an early sowing. The problem is that in spring it is important that autumn sown plants be kept cooler than spring sown ones and this is not always easy to manage. Otherwise, with the start the autumn sown ones have had, they end up too tall and leggy. Apart from geraniums, begonia and campanula, antirrhinums, dianthus, gazania and penstemon might be worth trying and they can go outside that little bit earlier than the less hardy types.

**Overwintering**

With more and more people putting conservatories on their houses fewer have to resort to the windowsill for germinating seeds, which is just as well as it is not an ideal spot. There are a couple of problems. The light is weak and comes only from one side, and the temperature fluctuates unpredictably. And of course there are not all that many windowsills in the house where you want to have seed trays lined up — they are not very elegant and in most rooms you would probably rather have pot plants.

**On the Windowsill**

*Figure 1.19 The geraniums on the left were raised on a windowsill, those on the right in a greenhouse. Both were sown at the same time*

Although the plants do get very leggy and drawn, and often rather susceptible to disease, they will recover in the end, but there is often a delay of some weeks before the display in the garden is comparable with greenhouse grown plants.

There are one or two things you can do to minimise the problems. It helps to lay some kitchen foil on the windowsill before setting out the pots. This reflects light up around the plants. Another trick, if you are using trays, is to lift them up above the level of the window frame. The frame can cut out a surprising amount of light and lifting trays up just a couple of inches can make a big difference. A propagator can help too, preferably a thermostatically controlled one which will equalise temperatures so avoiding a cold spell during the night when the central heating goes off. Once the seedlings come through you will probably need to turn the pots regularly to avoid the seedlings getting too drawn, so contraptions with a vertical sheet of foil between the plants and the room are less than ideal — they make it difficult actually to get at them.

**Hardening Off**    Before plants are set out in the garden it is vital to acclimatise them to their new growing conditions. If plants are moved from a cosy greenhouse straight into the garden they suffer from chilly nights, cold wind and sun scorch. They are therefore moved into a cold frame as an intermediate stage. At first the frame is kept closed and then over the next couple of weeks it is opened first on good days then on colder days too; then at night as well

*Figure 1.20 A cold frame, though not necessarily one this size, is invaluable for hardening off plants*

until by planting out time they have been standing in open frames for a few days. If you do not have a frame then you can steadily increase the ventilation on the greenhouse before moving the plants to a sheltered spot outside and then into the open. Plants raised on the windowsill can be moved first into a colder room, then into the porch, then into a cosy corner of the garden before they go into their final homes.

## Buying Plants

Few gardeners are able to raise all their own plants themselves. There is not usually the space or time and accidents happen which necessitate buying plants. Both retail and mail order sources are important. Plants are available in garden centres, markets, garden shops, road-side wagons, at gates, in department stores, in DIY and gardening multiples but, generally speaking, the specialist garden centre or nursery is to be preferred. The reason is that you will probably find a better choice, better labelling, the plants will be looked after more carefully and you may even get some good advice from the staff. Prices may or may not be higher than other outlets. The only thing to be said about price is that markets are usually the cheapest but the plants are usually less well advanced, even though the quality may be good for their size.

A large variety of containers is available. The old wooden trays are still seen in some markets and the plants still cut off in strips with an old kitchen knife. Many are sold in flimsy, half sized seed trays usually of 20 or 24 plants, especially in markets. The polystyrene strip, often linked in a block of four or five, is also very common but the number of plants in each strip varies dramatically — you will need to count up and relate the content to the price. An increasing number of plants is now grown in pots, or packs of about six thin plastic pots linked together. The price is high, but for small quantities of plants, especially for containers, the quality is worth it.

Features to look for when buying bedding plants are:

(1) Plants of balanced shape, reflecting the natural characteristics of the variety. Reject leggy plants.
(2) Healthy green foliage with plenty of leaves low down on the plant.

(3) Lack of pests and diseases.

(4) Good sized, bushy plants, which have had space to develop.

(5) Not too many roots penetrating the drainage holes.

(6) Moist compost with no wilted plants.

When it comes to buying by mail order, you obviously do not have a chance to inspect the plants, but generally speaking the standard is high and you can often buy varieties unavailable at the garden centre. Plants are also available in a seedling form not seen in the garden centre and provide an especially useful buy for the home gardener. Small trays of seedlings at the pricking out stage, with between 100 and 250 seedlings per tray, come in a limited range of types, begonias and petunias being the most popular. Both these flowers have very small seeds which not everyone finds easy to sow. For most people this number of seedlings will be more than is needed but neighbours and friends can get together to order a number of types and then split them between the group.

Germinated seedlings of the more special varieties are sometimes available by mail order in petri dishes, and these are usually expensive but may be of varieties available nowhere else. The pricking out stage is especially difficult as the environment on the greenhouse bench is rather different from that in the petri dish but with a little care the transfer is not difficult to manage.

*Figure 1.21
Pot-ready plants in
Jiffy 7 pots come
packed in a sturdy
polystyrene case, to
prevent damage*

Geraniums, fuchsias and tuberous begonias are available from seed companies in Jiffy 7 pots and this is a very convenient way of buying them. Ten plants come in a special polystyrene box at the ready-for-potting stage and, although not cheap, they take all the worry out of raising plants. Make sure they are moist when you pot them and do not firm the compost. Specialist nurseries list an enormous range of vegetatively propagated geraniums and fuchsias and plants from them are usually dispatched as rooted cuttings in soil blocks. Again they are ready for potting and some may even be in bud. They can go on the windowsill until they are ready for planting out.

# CHAPTER 2
# *Cultivation*

## Soil Preparation

Annuals and bedding plants only have a short time in which to get established, grow and do their stuff so it is important that they have as good a soil as possible. As I have said, the myth about annuals requiring a dry, poor soil really is a myth. In general, annuals and bedding plants need a soil that is sufficiently moisture-retentive to carry the plants through dry spells without constant watering, with sufficient plant food to satisfy their nutritional requirements. In new soils these requirments can be met by, at the simplest, forking in peat and raking in a general fertiliser such as Growmore, and this can work very well. But to be more specific. In beds or borders that have a constant succession of annual or bedding plants, you will be removing quite a volume of plant material twice each season. You have to put something back to make good the loss and the timing depends on your soil and how the job fits in with your other garden tasks. Personally, I tend to add the muck in the autumn because I have fewer other things to do at that time of year compared with the spring. So when the summer plants go on the compost heap, the compost made from the spring plants goes in. If you do not have the facilities to make compost, or have too many other demands on what you have, use peat, mushroom compost, manure — anything will do. If it is very well rotted it can be forked in; if not it is best left in a layer in the traditional manner.

In late spring, when the spring plants come out, a simple forking through followed by a dressing of Growmore at

2–4oz per sq yd (60–120g per sq m) depending on the inherent fertility is about right. When the fertility of the soil has built up well, you will be able to dispense with the annual addition of organic matter and every two, or maybe three, years will do. Where the area given over to annuals will eventually be used for more permanent plants, often the case in new gardens, generous additions of organic matter should take place every year, or even twice a year in spring and autumn. Indeed one of the best ways of increasing the long-term fertility of the soil in preparation for shrubs and trees is to grow annuals for a few years while you deal with other areas of the garden, or the house if it is a new one, and to put plenty of muck in between plantings.

In pockets between permanent plants, be they shrubs or border perennials, it is more difficult to dig and so forking in a material like peat or leafmould which is fine and well rotted should be used as it is easier to deal with in a confined space.

## Planting

When it comes to planting time you may or may not have a clear idea of how the plants are to be grouped in the beds. Either way, get all the plants out of the frame or greenhouse and down to the site and then water them well. Have a cup of tea while the water sinks in, or get the bed raked down and the fertiliser on. Then set the plants out on the bed. It pays to get all the plants for a specific area laid in place on the soil before planting any of them as this gives you the opportunity to move things around, to finalise the arrangement before actually getting to work with the trowel. Go so

*Figure 2.1 When planting bulbs and wallflowers together, put the bulbs in after the wallflowers, otherwise you might risk damaging the bulbs*

*Figure 2.2 Set out
summer bedding
plants to make sure
the spacing is right
before planting*

*Figure 2.3 Try to
retain as much soil
as possible on the
roots of plants
when you pull them
apart for planting*

far as to cut individual plants out of the trays and lay them
out, moving them to just the right spots. It can be a great
help when doing this to use your trowel as a measure. If
you know, for example, that the blade is 6in (15cm) long,
then you can be quite precise as to planting distance

31

without recourse to measuring rods. When it comes to getting the plant in the ground, Alan Cook, my old boss at Kew, used to say that you need just three moves with a trowel to make a hole big enough for a plant from a 3in (7.5cm) pot — and it is true, but it takes practice and the soil has to be good.

After planting, firm in well with two fingers on either side of the plant, and then water in. There is a temptation to put the sprinkler on and leave it for an hour if you have planted a lot but a far better idea is to use a can and give them a feed at the same time. Especially if plants have come from trays, they will have lost a lot of root and so a good drink of liquid feed straight after planting will help them recover quickly.

## Pinching, Pruning and Dead-heading

One of the aims of plant breeders in recent years has been to create varieties which are naturally bushy and so do not need pinching. However, less than ideal growing condi-

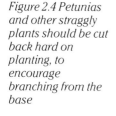

*Figure 2.4 Petunias and other straggly plants should be cut back hard on planting, to encourage branching from the base*

tions can cause nicotianas, salvias, antirrhinums and petunias to run up leggily and then it pays to take the tops out when planting.

Some plants need attention during the season if they are to give their best and this applies to container plants more than others. Of course the discreet removal of stray shoots in a garden display can make all the difference, as can the careful encouragement of shoots to grow in certain directions, but ivy-leaved geraniums and helichrysums need thinning or cutting back occasionally to keep the balance right. Think about it and do not be afraid.

Dead-heading is probably the second most important job after watering, once the plants are in. Obviously very small-flowered plants like lobelia are impossible to dead-head. If they could be dead-headed they might do rather better at the end of the season. But the simple fact is that to prevent plants putting a large proportion of their energy into fattening up seeds at the expense of producing more flowers — the dead flowers have to come off. There is also another reason, the ripening seed heads produce hormones which are spread to the rest of the plant to depress the production of more flowers. Some plants, mainly triploids like the Afro-French marigolds and begonias like 'Pink Avalanche', do not produce seed so dead-heading is unnecessary although the plants may look better. With everything else it pays off handsomely. Most gardeners spend a lot of time in the garden in spring and summer — otherwise what is the point of all this gardening? So when wandering around feeling pleased with your handiwork, always pick off the faded heads.

And before going on holiday for a couple of weeks remove even the fully open, unfaded flowers so that when you come back after a fortnight in the sun, your plants will be in full flower. No, I do not suggest that you spend a week on your knees picking off perfectly good flowers all over the garden. But tubs and baskets certainly benefit from the treatment.

## Weeding

The hoe is the normal implement for such work and will serve the purpose very nicely, but a few refinements are in order. There will not be a great deal of clearance between the edges of most hoes and the plants from around which the weeds are being removed. So to avoid damaging foliage it is sensible to use a Paxton hoe (from Bulldog Tools) rather than a Dutch hoe — the Paxton has a single central strut fixing blade to handle while a Dutch hoe has a strut at each edge and these side struts tend to knock low foliage. Another useful trick, to avoid accidentally slicing off the stems of young plants, is to file the corners of the hoe from points to curves so that if you do touch a plant, the stem will slide around without damage. These precautions are unnecessary with plants like sprouts which are tougher and set at wider spacing but for closer and more delicate plants it does pay off.

Weedkillers can be of use too, although a weed preventer is more useful. Propachlor is a short-term weed preventer which will kill weed seeds as they germinate for six to eight weeks. It comes in a granular form and is scattered on clean ground straight from the container. It is only on salvias that it cannot be used and after the active period it breaks down in the soil. By that time the plants should have grown sufficiently to smother most of the weeds that come through. The same applies to hoeing — eventually the foliage cover does the job.

## Watering and Feeding

Watering is not always necessary in rainy summers, but in some years a good sprinkler (and a licence for it) is the saviour of many fine schemes. Multiflora geraniums, nemesias, lobelia and mimulus are just some of the plants especially likely to suffer in hot dry spells. Leave the sprinkler on for a good long time and then give the plants plenty of time to get their roots well down before giving them another dose. It is no good standing there with a hose pipe; you will get fed up before enough water has gone on.

If the soil is prepared well, the feed the plants get when watered in should be the last they need. Only plants in containers will need more, and that is covered in the appropriate chapter.

## Staking

For hardy annuals in particular this can be an important element of a successful display as quite short plants sometimes need support, even if only to prevent their flopping on to the lawn. The traditional method is to use brushwood but this is not widely available these days, especially hazel sticks which are the best. The great thing about hazel is that the branches come in flat sprays of fan like growth. This means that each piece of brushwood supports a lot of plants for the number of twigs it contains. The flat sprays can go round the edges of groups and diagonally across the middle making very efficient supports. They should be about two-thirds of the eventual height of the plants and the tops can be broken horizontally across the tops if they are too tall. Unfortunately, the birch twigs which are more common are besom-like in their arrangement and altogether less efficient.

Canes and string make a more formal system. A cane at each corner of a rectangular group or five or six around a more irregular shape supports strings run from cane to cane and looped round each. The strings are also run across the tops of the plants and are looped around each other where they cross. This is less elegant but if done at an early stage is fairly unobtrusive. The advisability of setting supports up early cannot be over-emphasised. There is nothing more unattractive than a collapsed group in need of a rescue.

Another method, adapted from commercial cut flower practice, involves plastic pea and bean netting stretched horizontally over the plants and supported at the sides and in the central area too. After germination and thinning, weeds are removed, a weed preventer applied and the netting stretched across. With careful fixing, it can be made lower at the front than the back. The plants grow through the netting, which has a wide mesh, and are supported effectively.

# CHAPTER 3
# *Pests and Diseases*

There are two approaches to dealing with pests and diseases — preventive and curative. In general, preventive measures are by far the most effective although when it comes to chemical control this may not be so. The most important preventive measures are good garden and greenhouse hygiene and careful inspection of incoming plant material. Care for your plants — a well-grown healthy plant is less likely to suffer than a weak straggly one.

Garden and greenhouse hygiene is one of the most important preventive measures. Remove dead leaves and flowers from the greenhouse, do not leave dirty pots lying about, remove weeds that may act as alternative hosts and scrub down the greenhouse well each year. Only buy plants from reputable suppliers and only buy healthy plants. If buying by mail order, inspect plants carefully when they arrive. I hate to say it, but inspect plants given by friends and neighbours particularly carefully.

Make sure that plants are happy and in good general condition and they will then be more resistant to attack. Grow them in the conditions that suit them, water them in the way they like, feed them according to their needs and the fertility of the soil.

Chemical sprays can also be used preventively but I am not inclined to use chemicals unless I have to, so tend not to spray unless I have evidence of attack. There are exceptions, and calendula are almost guaranteed to get mildew and hollyhocks to get rust, so both should be sprayed regularly as a precaution. Otherwise, wait. But inspect plants thoroughly and regularly so that you spot problems at the first sign. This does not mean making regular pest hunting forays every Monday and Thursday, it means keeping your eyes open whenever you are in the garden.

Curative measures are not just chemical in nature; if you spot the problems early you can pick off the offending beasts and pop them on the fire. I have a friend who picks up every slug and snail she finds and throws them over the wall on to the waste ground on the other side. This works fairly well, although they have sometimes been spotted crossing the top of the wall back to her side. If chemical control is necessary do it thoroughly. Spray the undersides of the leaves as well as the tops.

**Aphids**

Aphids in a wide variety of colours can attack just about every plant at every stage of growth. A specific aphid killer such as pirimicarb is the wisest choice because it kills no other insects and spares many helpful predators. In the greenhouse, a fumigant is often more effective.

**Botrytis**

Very troublesome in wet seasons. A noticeably fluffy, grey mould appears on flowers in particular but also on buds, leaves and stems. African marigolds, double zinnias, double dahlias and double geraniums are especially susceptible as the flowers collect water and then turn an unsightly brown. Pick off the affected flowers to prevent spread and spray with a systemic fungicide such as benomyl or thiophanate-methyl.

**Blackleg**

Blackening of the stem just at and above soil level, usually in pot-grown plants. Geraniums are especially prone, in particular if kept too wet in the winter. Burn infected plants and water others with a copper fungicide.

**Capsids**

Small irregular holes first appear in the leaves around the growing point and expand as the leaves expand. The tiny green insects attack the very youngest leaves in the tip. A systemic insecticide such as pirimophos-methyl will deal with them.

**Caterpillars**

A variety of caterpillars in various colours and sizes can be troublesome, especially leaf rolling types, but they can

38

usually be picked off container-grown plants. In beds, they can be dealt with by a thorough spray with an insecticide containing derris.

**Cats**

However much you love cats, they will not be popular if they dig up newly sown annuals. In small areas one of the best preventives, although it sounds rather crude, is to put sticks with sharpened tips in the soil where the annuals have been sown. . . I have tried this and am convinced that it is not the pointed tips that leads to success but the inconvenience, and the difficulty of manoeuvring in such a situation. Pepper is useless after dew or a shower and having tried all the pellets and powders available I can testify that the cats continue as if they were not there, even squatting immediately over groups of pellets. A hose pipe or bucket of cold water will deal with the neighbour's cat and black cotton stretched across the area will also put them off.

**Damping Off**

Germination may appear poor, young seedlings collapse from soil level, older seedlings may appear stunted, or may wilt unexpectedly, they may go yellow and a general unevenness of growth will be apparent. Unfortunately there is always the temptation, when wilting is seen, to water more often — the worst possible course. Damp compost is fatal. But there are a number of precautions that can be taken to prevent the occurrence of the disease and minimise its effects once it has arrived.

(1) Clean the greenhouse thoroughly, especially benches, glass and glazing bars. Commercial growers use formaldehyde but for home gardeners this is a very dangerous material and it is generally wiser to avoid its use — if only because a large amount of protective clothing is needed to do the job safely. A strong household disinfectant is the best bet.
(2) If you use capillary matting to help with the watering, use fresh material each year for the antirrhinums, begonias and seedlings in general. Disease can spread from contaminated matting.
(3) Use new pots and boxes for sowing and for pricking out. Small pots or packs will help minimise the spread of the disease if it appears.

(4) Use fresh compost, preferably not a peat only type. The extra drainage of composts with added vermiculite, perlite or grit will help to keep the compost sweet and open.

(5) Drench with a fungicide (such as a copper fungicide) before sowing.

(6) Sow thinly in small pots. If growing a lot of the same variety sow in two small pots rather than one tray, and if one is stricken only grow on the healthy one.

(7) Never keep the propagator unnecessarily warm and always remove condensation.

(8) Only prick out perfectly healthy seedlings and again make sure to use the right compost.

**Earwigs**

Earwigs are infuriating beasts which cause problems just at the moment when you think you have achieved your goal of the most beautiful blooms in the street. They attack a number of plants but are especially troublesome on the flowers of double dahlias and chrysanthemums late in the season. Earwigs are nocturnal and retreat into dark, damp spots during the day, hence the traditional approach of inverting clay pots stuffed with hay on the stake. The earwigs retire there at dawn and the gardener can then empty them on the bonfire. A more modern alternative is to use an insecticide based on permethrin.

**Froghopper (Cuckoo spit)**

This is not usually a fatal problem, but is unsightly and debilitating. Inside the foam is a small, green, sap-sucking insect. A systemic insecticide such as dimethoate is the answer.

**Leaf Miner**

A tiny grub burrows just under the upper surface of the leaf causing translucent tunnels. This is especially troublesome on chrysanthemums and other members of the daisy family, including some weeds like groundsel. Pick off the offending leaves as soon as the first tunnels are seen and that may be enough. Otherwise, spray with an insecticide that penetrates the leaf or is systemic, such as pirimiphos-methyl or dimethoate.

Most common in damp seasons, leaves turn yellow with grey mould on the underside; wallflowers are especially prone. Spray with a systemic fungicide such as benomyl. **Downy Mildew**

In contrast to downy mildew, powdery mildew is more common in dry seasons, especially when the soil is dry. Larkspur, calendula, cornflower and verbena can all suffer badly although some varieties are more resistant than others. A systemic fungicide such as propiconazole will cure the problem but regular sprays are necessary. **Powdery Mildew**

Possibly the most infuriating of pests as they are so elusive — the damage is apparent enough, but catching the beasts. . . There is no foolproof answer and most of the suggested remedies have been successful for someone. Traps and poisonous smokes will certainly kill moles if used carefully, but many gardeners would rather scare them off and this is less easy. Early risers can creep into the garden in bedroom slippers at around dawn armed with two spades and a biscuit tin. Most mole tunnels are dug at dawn so when you see evidence of movement, dig in one spade to cut off the mole's retreat and use the other to dig him (or her) up. Pop the beast into the tin and drive to a spot where the mole can be released. They can swim, by the way, so do not expect a river to stop them returning. **Moles**

Especially common in dry seasons, and in dry parts of the garden such as against south-facing walls. The leaves first turn speckly and then yellow and a fine webbing appears. Close inspection reveals large numbers of very tiny red or orange mites which suck sap and cause dramatic debilitation. It is more common in the house and greenhouse and often gets a hold on plants before they go outside. Fumigation of the greenhouse together with the prevention of a dry atmosphere are very helpful and outside derris is the safest control spray. **Red Spider Mite**

This is covered in detail under althaea and antirrhinum, but is very difficult to control except with propiconazole. **Rust**

**Slugs and Snails**

Slugs and snails can be very damaging in wet periods, especially to young and succulent shoots. Liquid preparations which are watered on are the least likely to create problems for wildlife and a gel preparation is also good in this respect. Pellets last longest in wet conditions and should be used very sparingly, modern mini-pellets containing methiocarb should be used singly about 6in (15cm) apart. However, traps using stale beer or milk and water combine effectiveness with almost total harmlessness to other forms of life.

**Virus**

Virus is less of a problem amongst bedding plants and annuals than amongst most groups because so many are annual and it is exceptional for a virus to be transmitted through seed. Named geraniums propagated from cuttings are the worst sufferers. To keep all plants virus-free, keep good control of the insects which carry the virus, especially aphids.

# CHAPTER 4
# *Display*

The whole point of growing bedding plants and annuals is to bring colour to the garden — though form should not be ignored. Everyone's perception of what makes an appealing display is different so although I urge you to try some of the ideas I suggest, do not be afraid to try ideas of your own. Pinch ideas you see in other people's gardens and adapt them to your own. Go to Kew and Christopher Lloyd's garden at Great Dixter in Sussex and look at the beds in the care of the local authorities in Cambridge and Brighton.

## Mixtures

Beds and borders given over entirely to annuals and bedding plants enable you to create dramatic and colourful displays. In particular it is in these situations that mixtures of colours really work. You will soon gather that I am not a great fan of the 'mixed colours' approach to choosing varieties and in borders with other plants they can sit very uneasily. In beds alone, they often work well and have the interesting capacity of appearing to change as you walk closer. From some way off a mixture can almost appear as a single bright colour, rather in the way that a colour photograph in a magazine is made up of dots in different colours which from reading distance form a uniform tint. As you approach, the individual dots become more apparent and you can look at the way the different coloured plants interrelate. And as you come up close, the relationship between the different flowers dominates.

There are problems with mixtures. Unless you buy a 'formula mixture' there will have been little control over the balance of the different colours included. A formula mixture is made up of a constant proportion of the different colours so that if you prick out every little seedling you should get the balance of colours that the seed company has decided makes the best display. It was once necessary to prick out both vigorous and weak seedlings as some-times flower colours are associated with particular degrees of vigour. Modern plant breeders are eliminating this distinction but with so many good old varieties still around the advice holds. There is still a problem, though, because although the colours may all be there in the right proportions, there is no telling where they are going to appear in the bed when you plant them out. Most home gardeners do not have the facility to ensure that their plants have opened the first flowers when they are planted out, and indeed this is not necessarily a good thing, so you could easily plant a number of plants of the same colour next to each other — you just do not know. In small beds this can be especially infuriating. Some mixtures may contain ten or twelve colours and you may end up with a bed from which two or three are missing.

There are, however, two plants which work especially well in mixtures on their own, fibrous rooted begonias and impatiens. In both, there is an enormous range of shades. In begonias the shades are in a fairly tight limit from dark red through pink to white with picotees, but there is also the option of dark red foliage, green foliage and various intermediate shades. 'Lucia' has 13 combinations of varying colours and dark or green foliage. In impatiens, 'Super Elfin Mixed' has eleven colours, 'Cleopatra' has 16 while there is also a mixture solely of softer shades, 'Pastel Mixed'. In some ways the imps are a little less satisfactory as mixtures as they tend to spread out giving large blobs of colour from a distance rather than the sparkling colours of begonias; the new 'Mini' series may be better in this respect.

One thing to avoid with some determination is using mixtures of two different plants next to, or amongst, each other. You sometimes see this in spring with wallflowers and tulips and it makes a very confused bed. One of the most unpleasant combinations I have ever seen was made up of a carpet of mixed salvias with blocks of mixed 'Madame Butterfly' antirrhinums scattered through them —

such a waste. What does work well is to combine a mixture with one single colour in the form of foliage or flowers. An ideal companion for the 'Pastel Mixed' impatiens would be a grey foliage plant, and a border of white petunias in front of mixed geraniums would look pretty good.

To avoid the big problem of unpredictability, one final thought on mixtures is that there is nothing to stop you making your own mixtures. Try antirrhinums for example — 'Black Prince' with deep crimson flowers and bronze leaves looks well with 'White Wonder' with pure white flowers and fresh green foliage. If the flowers are too dark for you, 'Purple King', mauve with an orange throat, is a good bet too. You will need to judge the proportions carefully and in this case use rather less than half of the white but add a white edging in fibrous begonias or alyssum.

## Informal Arrangements

In the modern garden, the informal use of bedding plants and annuals will be the most successful; in few gardens do the stripes or blocks of old fashioned parks' bedding have a place. Rather, an arrangement based on the layout of the herbaceous or mixed border, popularised by gardeners such as Christopher Lloyd, Graham Thomas and Margery Fish, will be more suitable. In this system plants are used in informal blocks, sometimes merging one into another, thoughtfully organised so that the plants relate to their neighbours and near neighbours in terms of flower colour and form, foliage colour and form, general plant habit, height and flowering period. The situation may vary from a long border with a good background of timber fence or hedge, or a similar border with no more backing than a chain link fence; it may be a narrow border alongside a path, a space in a large mixed border, a corner bed by a small patio or a round bed in the middle of the front garden lawn. Plantings of this sort also give you the option of planting out tender perennials like palms and standard fuchsias which are kept from year to year.

## Carpets

In parks or on traffic roundabouts you will sometimes see vast carpets of colour, often marigolds or geraniums, where one variety in a single colour is used to cover a wide area. This makes a very dull display in the garden at home. You can plant a whole bed full of 'Inca Orange' marigolds and it will be very bright and colourful all summer — but it will also be very boring. This type of display is well suited to roadsides where fleeting peripheral vision is all that is brought to bear in its appreciation. But when you may spend half an hour at a time gazing at it out of the kitchen window, while cooking or washing up, something more satisfying is in order.

## Left Overs

Whether or not we plan our displays carefully in advance, to give us a precise guide to the plants required — there are always some left over. These sometimes find themselves squeezed in amongst their siblings so that the plants are overcrowded but I tend to put them all together in one bed. This is a splendid way to create surprises, try out new and outrageous associations and get ideas for plantings on a larger scale in future years.

## Temporary Plantings

When shrubs and herbaceous perennials are first planted they are small and do not take up much space so something is needed to fill the gaps. Annuals are especially useful in this respect as they are not expensive, they can be sown where they are to flower and the twice-yearly gaps between plantings give a vital opportunity to get some organic matter into the soil, which is essential for the long-term wellbeing of permanent plants. Be prepared to cut back any plants which encroach on those which have a long-term future. Even young shoots low down, which are easily shaded out by vigorous annuals, should be protected by trimming back the annuals, as these low shoots create the bushiness that is so important to an elegant shape in later years. Use predominantly low plants, restricting taller types to the spaces between your permanent plantings so that there is no shading or competition. As the permanent subjects grow, the area given over to annuals can be reduced until eventually there are only some carefully spaced gaps which are intended for annuals or bedding in the long term.

# Planting in Perennial Borders

There are two ways of approaching this — gaps can be left specifically to take a changing display of temporary plants, or annuals and biennials can be slotted in amongst permanent subjects, sometimes right in the middle of clumps of border perennials, where no large space has been left. Both these approaches are suited to those with relatively small gardens where there is no space to give an area over specifically to temporary plantings. And small gardens are probably less likely to have spacious propagation facilities and so there is less scope to produce large quantities of plants.

If you choose to leave some spaces you can fill them in two ways. You can choose plants to associate with the flowering times of the surrounding plants or you can use surrounding plants simply as a background, intentionally not co-ordinating flower times. Using a summer-flowering shrub such as the rose 'Nevada,' a creamy white shrub rose with a touch of pink, you could incorporate this into a summer scheme by filling an area in front with silver foliage such as *Cineraria* 'Silverdust', a pale pink petunia such as 'Chiffon Magic' which picks up the hint of pink in the rose flowers or a darker shade could be used such as one of the taller fibrous begonias like 'Rusher Rose'. If you prefer a contrast a dark lobelia like 'Crystal Palace' might do the trick, although it is a little short, or the purple *Salvia* 'Laser Purple'. Or, another salvia, the taller, slimmer 'Victoria' could be used at the side and will contrast well with the tumbling rose sprays plus the begonia or a pink impatiens like 'Super Elfin Blush'. If you do nothing else, you could plant an *Ipomoea* 'Heavenly Blue' at the base to climb through.

In spring you will not have the flowers to think about, just the fresh young foliage so you can change the colours dramatically and go for 'Blood Red' wallflowers fronted by 'White Floral Dance' pansies. The colours of these two could be altered each year but I would be inclined to use a tallish wallflower directly in front of the rose to help cut off the view of the bare stems; later the rose's own foliage will do the job.

If, on the other hand, you have an area of winter-flowering shrubs like *Viburnum tinus*, then you can ignore the flowers entirely. The dark viburnum foliage would provide a good background for pale colours and a group of

'Ivory White' wallflowers might look good, or lemon tulips like 'Reforma' with *Myosotis* 'Royal Blue'. Later the options are endless but should include sweet peas or other climbers to add colour high up amongst what would otherwise be dull green.

When it comes to slipping a few plants in the borders without making specific spaces for them, they will often look after themselves. A white cosmos like 'Purity', once planted near some strident phlox or delphiniums to calm them down a little, will always be around if left to seed each year and seedlings may just need moving occasionally or removing from awkward spots if they do not look like doing their job for you. The same applies to biennials like cotton thistles and myosotis. Perhaps the most useful group in this respect are climbers which can usually be slotted into the most crowded borders to good effect. Sweet peas obviously work, as long as they are watered and fed well when first set and 'White Leamington' is splendid trained through the purple smoke bush (*Cotinus coggygria* 'Royal Purple'), the soft pink 'Southborne' can go through a summer ceanothus like 'Autumnal Blue'. I am sure you get the idea. Ipomoeas are used in this way too, especially as a shrub can often give them useful shelter from cold winds. More woody plants like eccremocarpus tend to produce growth which is a little too heavy and they are best trained up a fence in the first instance and then allowed to stray.

# CHAPTER 5
## *Colour*

Colour is surely the prime consideration in almost all forms of flower gardening and also the one about which there is least agreement. I will suggest an overall approach to the subject and where I do not actually say so you will doubtless learn where my tastes lie. And remember that it is not just the flowers that provide colour, foliage too, be it green, yellow, purple or bronze should be treated as important.

There are three main approaches. You can aim for a contrast of colours like the familiar red salvias, 'Red Riches' is the one to go for, blue *Lobelia* 'Emperor William' and white *Alyssum* 'Carpet of Snow'. There is no doubt that it will be noticed. Then there is an approach which Gertrude Jekyll used greatly, planting different shades of the same colour together. The Sunburst display of marigolds at the Liverpool Garden Festival was a prime example with varieties ranging in colour from lemon yellow to rich bronze. Finally you can match strengths of colour. Pale blue, soft pink and grey is a familiar example — *Ageratum* 'Ocean', *Petunia* 'Birthday Celebration' and the grey foliage of *Pyrethrum ptarmicaeflorum*.

## Contrasting Colours

Contrast certainly provides the most arresting spectacles but also those which pall the soonest. You can add variety by using unfamiliar plants in these colours; an alternative to the red, white and blue given above would be *Lavatera* 'Mont Blanc', *Geranium* 'Solo' and *Pansy* 'Jumbo Blue'.

Other examples include:

**Spring**

*Anthemis cupaniana* (from cuttings) (white flowers, silver foliage)
*Iris* 'Purple Sensation'

*Myosotis* 'Royal Blue'
*Tulipa* 'West Point' (yellow)

*Myosotis* 'Blue Ball'
*Narcissus* 'Baby Moon' (bright yellow)

*Myosotis* 'Royal Blue'
*Narcissus* 'Lintie' (yellow and orange)

*Cheiranthus* 'Ruby Gem'
*Tulipa* 'West Point' (yellow)

*Cheiranthus* 'Primrose Monarch' or 'Cloth of Gold'
*Tulipa* 'Queen of the Night' (purplish black)

*Lunaria annua* (purple)
*Tulipa* 'Golden Appledorn' (gold)

*Lunaria annua* 'Alba'
*Tulipa* 'Queen of the Night'

*Bellis* 'Kito' (pink)
*Scilla* 'Spring Beauty' (dark blue)

Polyanthus 'Pacific Red'
*Narcissus* 'Mount Hood' (white)
*Bellis* 'White Carpet'

*Cheiranthus* 'Blood Red'
*Tulipa* 'Purissima' (white)

*Anthemis cupaniana*
*Tulipa* 'Burgundy Lace'

*Myosotis* 'Blue Ball'
*Tulipa* 'Red Riding Hood'

*Cheiranthus* 'Blood Red'
*Tulipa* 'White Triumphator' or 'Niphetos' (both white)

*Cheiranthus* 'Blood Red'
Pansy 'White Floral Dance'

*Aubrieta* 'Ruby Cascade'
*Narcissus* 'Tête à Tête (dwarf yellow)

*Myosotis* 'Blue Bouquet'
*Cheiranthus* 'Allegretto' (orange)

Pansy 'Prince Henry' (yellow)
*Muscari* 'Blue Spike' (mid blue)

**Summer**

*Salvia* 'Red Riches'
*Calceolaria* 'Midas' (yellow)

*Nicotiana* 'Domino White'
*Convolvulus* 'Royal Ensign' (blue and white)

*Eucalyptus globulus*
(silver foliage)
*Hibiscus* 'Coppertone'

Ruby Chard (red stems)
*Cineraria* 'Silverdust'

*Perilla atropurpurea*
'Laciniata' (purple
foliage)
*Petunia* 'Recoverer White'

Cabbage 'Red Drumhead'
*Helichrysum* 'Limelight'
(from cuttings) (yellowy
green foliage)

Brussel Sprout 'Rubine'
*Tropaeolum canariense*
(yellow)

Basil 'Dark Opal'
*Pyrethrum* 'Golden Ball'

*Kochia trichophylla* (green
foliage)
Geranium 'Solo' (red)
*Cineraria* 'Silverdust'

*Salvia patens* (blue)
Geranium 'Solo'
*Tagetes* 'Lemon Gem'

Marigold 'Doubloon'
*Antirrhinum* 'Purple King'
*Cineraria* 'Diamond'
(silver foliage)

*Begonia* 'Danica Red'
*Pyrethrum* 'Golden Fleece'

*Dianthus* 'Crimson Charm'
*Cineraria* 'Cirrus' (silver
foliage)

*Cuphea miniata* 'Firefly'
*Lobelia* 'Sapphire'

*Salvia* 'Laser Purple'
*Tagetes* 'Red Seven Star'

*Ricinus* 'Impala' (purple
foliage)
*Dimorphotheca* 'Glistening
White'

*Antirrhinum* 'Black Prince'
*Antirrhinum* 'White
Wonder'
*Begonia* 'White Avalanche'

*Tagetes* 'Inca Orange'
*Lobelia* 'Mrs Clibran' (dark
blue)
*Cineraria* 'Silverdust'

*Anchusa* 'Blue Bird'
*Verbena* 'Blaze' (scarlet)
*Lavatera* 'Mont Blanc'

*Atriplex hortensis* 'Rubra'
*Cosmos* 'Purity'

Geranium 'Grenadier'
(red)
*Lobelia* 'Mrs Clibran'
*Cineraria* 'Silverdust'

*Perilla laciniata*
'Atropurpurea'
*Pyrethrum ptarmicae-
florum* (silver foliage)
Parsley 'Bravour'

Swiss Chard (white stems)
*Verbena* 'Blaze'

# Different Shades of the Same Colour

In the next category comes variation in shade within a narrow limit. The first example of this that struck me was at Kew, where *Nicotiana* 'Lime Green' was once used with the yellow *Petunia* 'Brass Band' and parsley 'Bravour'. The Sunburst display at Liverpool used a range of marigolds, especially the Afro-French triploid type and in the garden you could choose 'Red Seven Star', 'Mata Hari' and 'Solar Sulphur'. An interesting variation on this theme which I saw recently was to use four white-eyed geraniums together — 'Ringo Dolly', 'Bright Eyes', 'Hollywood Star' and 'Minuet' — and very effective it was too. Other ideas in this style include:

**Spring**

*Myosotis* 'Carmine King'
*Tulipa* 'China Pink'

*Valeriana phu* 'Aurea'
*Narcissus nanus* (yellow)

*Cheiranthus* 'Primrose Monarch'
Polyanthus 'Pacific Lemon'

*Lunaria annua* (purple)
*Cheiranthus* 'Ruby Gem'

Pansy 'Fire Dragon'
*Tulipa* 'Orange Bouquet'

*Cheiranthus* 'Orange Bedder'
Pansy 'Gypsy Dance' (dark orange)

*Cheiranthus* 'Blood Red'
*Tulipa* 'Orange Bouquet'

*Cheiranthus* 'Fire King'
*Tulipa* 'Golden Appledorn'

*Cheiranthus* 'Ruby Queen'
*Tulipa* 'Queen of the Night'

**Summer**

*Antirrhinum* 'White Wonder'
*Cineraria* 'Cirrus' (silver foliage)

*Delphinium* 'Rosamund'
*Lavatera* 'Silver Cup'
*Petunia* 'Birthday Celebration'
Geranium 'Gala Amaretto'
All shades of pink

Marigold 'Paprika' (hot orange)
Marigold 'Queen Sophie' (coppery red)
*Helichrysum* 'Hot Bikini' (reddish orange)

*Ricinus* 'Impala'
*Eccremocarpus scaber* (orange)

*Salvia* 'Argent' (white)
*Helichrysum petiolatum* (from cuttings) (silver foliage)
*Impatiens* 'Super Elfin White'

*Centaurea gymnocarpa* (silver foilage)
*Dimorphotheca* 'Glistening White'
*Helichrysum microphyllum* (from cuttings) (silver foliage)

*Perilla laciniata* 'Atropurpurea'
*Calceolaria* 'Midas'
*Coleus* 'Red Monarch'

*Lavatera* 'Mont Blanc'
*Cineraria* 'Silverdust'
*Helichrysum petiolatum*
*Petunia* 'Recoverer White'
Parsley 'Bravour'

*Zea japonica* (variegated foliage)
*Cosmos* 'Purity' (white)

*Eucalyptus globulus*
*Lavatera* 'Mont Blanc'
*Petunia* 'Recoverer White'
*Antirrhinum* 'Giant Forerunner White'

*Helichrysum* 'Marine'
*Salvia* 'Laser Purple'
*Petunia* 'Resisto Blue'

*Hibiscus* 'Coppertone'
*Sanvitalia procumbens* (yellow)

*Dianthus* 'Princess Scarlet'
Marigold 'Red Seven Star'

Marigold 'Inca Orange'
Marigold 'Suzie Wong' (yellow)

*Ricinus* 'Impala'
*Dahlia* 'Coltness Scarlet'
*Tagetes* 'Scarlet Sophie'

## Matched Strengths of Colour

The third group includes matched colours in a similar tonal range and is, partly, the pastel version of the red, white and blue contrast of the first section. This approach is typified by the 'Pastel Mixed' formula mixture of impatiens which includes a rose, coral, pale salmon and other pinks plus lilacs, pale blues and white. Other associations in this group to try include:

*Myosotis* 'Carmine King'
*Lunaria annua* 'Alba'

*Cheiranthus* 'Rose Queen'    **Spring**
*Myosotis* 'Blue Bird'

*Cheiranthus* 'Blood Red'
*Tulipa* 'Princess Irene'

Pansy 'Super Beaconsfield'
   (blue)
*Narcissus* 'Hawera' (lemon
   yellow)

*Cheiranthus* 'Lemon
   Bedder'
Pansy 'Super Beaconsfield'

Pansy 'Paper White'
*Lunaria annua*
*Cheiranthus* 'Ivory White'

*Aubrieta* 'Blue Cascade'
*Narcissus* 'Hawera'

Pansy 'Imperial Light Blue'
*Tagetes* 'Solar Sulphur'

**Summer**

Geranium 'Hollywood Star'
   (pink, white eye)
*Cineraria* 'Silverdust'
*Lobelia* 'Cambridge Blue'

*Eucalyptus globulus*
*Ipomoea* 'Heavenly Blue'

Aster 'Milady Blue'
*Tagetes* 'Solar Sulphur'
*Petunia* 'Sky Joy' (pale
   blue)

*Salvia farinacea* 'Victoria'
   (purplish blue)
*Centaurea gymnocarpa*
*Verbena* 'Derby Salmon
   Rose'
*Verbena venosa* (purple)
*Lobelia* 'Crystal Palace'
   (blue)

*Eucalyptus globulus*
*Lavatera* 'Mont Blanc'
*Salvia farinacea* 'Victoria'
*Petunia* 'Chiffon Magic'
   (pale pink)

*Agrostemma* 'Milas'
   (cerise)
*Ageratum* 'Blue Bouquet'
*Petunia* 'Chiffon Magic'

*Crepis rubra* and *Crepis
   rubra* 'Alba' (pink and
   white)
*Ageratum* 'Blue Danube'

*Felicia amelloides* (blue)
*Petunia* 'Brass Band'
   (yellow)

*Lavatera* 'Silver Cup'
   (pink)
*Dimorphotheca* 'Glistening
   White'

*Salvia farinacea* 'Victoria'
*Pyrethrum
   ptarmicaeflorum*
*Petunia* 'Rose Picotee'

Marigold 'Silva' (yellow)
*Petunia* 'Resisto Blue'

*Antirrhinum* 'Black Prince'
   (very dark crimson)
Marigold 'Solar Sulphur'
*Lobelia* 'Crystal Palace'

*Anoda* 'Opal Cup' (silvery lilac)
*Agrostemma githago*

*Nicotiana* 'Domino White'
Geranium 'Appleblossom Orbit'
*Lobelia* 'Cambridge Blue'

*Impatiens* 'Novette' (mixed)
*Viola* 'Baby Lucia' (blue)

# CHAPTER 6
# *Containers*

In recent years there has been a surge of interest in growing plants in all sorts of containers. In the last year or two, hanging baskets have been especially in vogue and plant breeders have even developed plants specifically for growing in baskets — *Campanula* 'Krystal Mixed' and *Coleus* 'Scarlet Poncho' spring to mind. For the sake of convenience when it comes to garden planning I have grouped containers under three headings — tubs and troughs, hanging baskets, and window boxes — although as far as the planting is concerned there are many similarities.

## Tubs and Troughs

Fortunately, since interest in growing in containers has been on the increase, the range of containers available has increased in parallel — both at the cheaper end of the market, where the quality of plastic and fibreglass tubs reflects more thoughtful design and the matching of design to the materials and in the more expensive ranges, where replicas of classical urns in reconstituted stone give quality in a more traditional vein. Timber has been used more frequently too, with kits available at reasonable prices. What all this has done is to reduce the necessity for the construction of the more curious of home-made containers. In particular, the horror of the car tyre tub, represented even at the RHS Garden at Wisley I am sorry to say, can be said to be on the decline along with the rusty wheelbarrow and the cut-down plastic bottle. This variety gives gardeners more opportunity to choose containers to match the style of their gardens and the style of their houses too, so glaring incongruities can be avoided. Classical urns do not usually suit modern bungalows while concrete dishes look out of place around Victorian properties.

Compost for containers is a problem, partly because if you have more than one or two containers, and especially if some are large, you can use an enormous amount. It pays to use fresh compost every year although not necessarily for every planting. Start with fresh compost when the summer plants go in at the end of May and when these plants come out spring plants can go straight in. Ideally the compost should be removed and broken up as it will be bound together by roots and the addition of some perlite will improve the drainage — so important in the winter months. If you can get good quality soil-based compost this is ideal but is expensive in large quantities and few gardeners have the facilities to sterilise the quantities of loam necessary for home mixing. When you think that a big urn may take a whole bag of compost, you will get some idea of the expense. It does pay to use a soil-based compost in small, lightweight plastic urns as these can be unstable if light, peat-based compost is used. The ideal is to make your own and this is described on page 17. If perlite is used then no more will be needed when spring plants go in. After one year it is sensible to remove all the compost and use it as a mulch on the garden, starting again with a fresh mix.

Watering is one of the most crucial factors in maintaining a tub or urn. Earthenware and, to a lesser extent, reconstituted stone tubs lose water through the sides and this can be minimised by lining the pot with polythene before filling with compost. Peat-based composts hold a lot of water and with the addition of drainage material can be re-moistened easily if they should dry out. If you have a number of troughs and tubs, maybe around a patio, then the simplest way to water is to use a semi-automatic system. This is especially useful if you often go away for weekends and also makes it easier for neighbours to look after your plants when you are on holiday. The fine tubing used in greenhouse watering is ideal and can be run unobtrusively amongst the plants with one nozzle in small pots, two or three in larger ones and two or three in troughs. Use adjustable nozzles to give you precise control of the amount of water given to each container. The tubing is connected to the mains and turned on when the compost is dry. There are also units available which can be attached to the water line to feed plants when they are watered.

Feeding is very important in containers, especially those using a peat-based compost. They do not retain plant foods in the way that soil-based composts do, and after six weeks most of the plant food will have been used by plants or washed out. Branded fertilisers for flowering plants in containers are now available but in general it pays to start with a balanced feed and switch to one with a high potash content but which contains the other foods as well. My approach is usually to water the plants in with a liquid feed and then start feeding about a month later with half-strength feed upping to full strength every week or ten days from the end of July for summer plants. Spring plants will not usually need feeding so often; feed them on planting, and then they can be left until they start to grow in the spring.

When it comes to planting the same ideas suggested for beds and border apply except that you are restricted in the size of plants you can grow; but you have the added advantage of trailing plants which are rarely grown in the open ground. But do not eschew large plants altogether. A single angelica, or red castor oil plant can be very dramatic and tubs of other plants can be grouped around it. You have two ways of approaching the planting. You can grow a different plant in each container and group them together for the best effect, moving some right out of the way into a secluded corner when they are not looking impressive or you can treat each individually and make them self-contained pictorial units.

If you want to create self-contained displays then you can use a mixture with a little floppiness about it, say petunias, although something taller in the middle would probably improve it. Two plants work very well in most cases and for the central plants there is an enormous range. Geraniums and begonias were voted favourites in a recent trial, but African marigolds, *Salvia farinacea*, rudbeckia, nicotiana, dwarf dahlias, fuchsias, as well as tuberous begonias and geraniums are all good although unfortunately lavateras are a little tall for most tubs. Around them you can use one trailer or semi-trailer or something more bushy plus a trailer. Petunias, ivy-leaved geraniums, trailing fuchsias, 'Pink Avalanche' fibrous begonias, trailing tuberous begonias are all trailers that can be useful to fill the gap between the edge and central planting as well as used to trail. Trailing lobelia, *Campanula isophylla, Helichrysum microphyllum* and lysimachias are more genuine trailers while most bedding plants can fill the intermediate role. Some suggested combinations for tubs and troughs are:

**Spring**

Pansy 'Universal Mixed'
*Tulipa* 'Purissima' (white)

*Cheiranthus* 'Orange
  Bedder'
Pansy 'Gypsy' (dark
  orange)

*Cheiranthus* 'Primrose
  Bedder'
Pansy 'Super Beaconsfield'
  (blue)

*Cheiranthus* 'Scarlet
  Bedder'
Pansy 'Floral Dance White'

*Cheiranthus* 'Scarlet
  Bedder'
Pansy 'Ruby Queen'

*Cheiranthus* 'Vulcan
  Improved' (dark orange)
*Viola* 'King of the Blacks'

Pansy 'Floral Dance White'
*Muscari* 'Blue Spike'

Pansy 'Floral Dance Violet'
*Narcissus* 'February Gold'

Pansy 'Ullswater'
*Hyacinth* 'L'Innocence'
  (white)

**Summer**

*Cuphea miniata* 'Firefly'
*Lobelia* 'Sapphire'

*Marigold* 'Solar Gold'
*Chrysanthemum
  multicaule* 'Gold Plate'

*Salvia farinacea* 'Victoria'
  (purple)
*Petunia* 'Birthday
  Celebration' (pink)

*Salvia farinacea* 'Victoria'
*Marigold* 'Suzie Wong'
  (yellow)

*Pyrethrum ptarmicae-
  florum*
*Begonia* 'Pink Avalanche'
*Lobelia* 'Blue Cascade'

Geranium 'Scarlet
  Diamond'
*Cineraria* 'Silverdust'
*Lobelia* 'Sapphire'

*Dahlia* 'Rigoletto Mixed'
*Petunia* 'Recoverer White'

*Nicotiana* 'Lime Green'
*Campanula* 'Krystal Mixed'
*Helichrysum petiolatum*
  (from cuttings)

*Nicotiana* 'Domino White'
*Begonia* 'Pink Avalanche'

*Acidanthera mureliae*
  (white)
*Petunia* 'Resisto Mixed'
*Centaurea gymnocarpa*

Geranium 'Appleblossom
  Orbit'
*Cineraria* 'Cirrus'
*Lobelia* 'Blue Cascade'

Geranium 'Solo'
*Petunia* 'Resisto Purple'

*Centaurea gymnocarpa*
*Impatiens* 'Accent Mixed'
  or 'Pastel Mixed'

## Hanging Baskets

Hanging baskets are much in vogue these days, although creating a good show from baskets is more difficult than with tubs and troughs. This is mainly because the amount of compost most baskets hold is very small and insufficient to sustain the number of plants needed to make a colourful show. Until recently wire baskets were almost universally used. They were lined with fresh moss which held the compost in place and looked attractive. This type of basket is very pretty at its best but there are problems. Getting good moss is difficult these days and it should not be gathered from woods. The baskets dry out very quickly as

*Figure 6.1
Traditional wire baskets lined with moss give you the opportunity to plant trailing plants through the sides*

water is lost easily through the sides and the thickness of moss required to make a good cover and stop the compost falling out is fairly substantial, reducing the volume of compost. If the basket gets dry, the moss will go brown and look unsightly. The advantage is that the moss makes an ideal background for young plants and the structure of the wire baskets allows you to put plants through the sides whenever you choose.

Alternative liners are now available and these are made of sponge, compressed wood fibre or plastic. None of these give you quite the freedom of choice as to where to place plants in the sides and with the plastic and wood fibre it is quite difficult to cut holes. A useful improvisation is to use carpet underlay or black capillary matting which can be snipped fairly easily in the positions where you want to put plants through. Plastic baskets are now common and many have the benefit of a small water reservoir — which is just as well as many are very shallow

*Figure 6.2 Modern plastic hanging baskets often come with a useful built-in water reservoir*

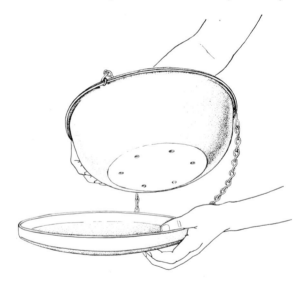

for their diameter. They lose no water through their sides of course, but with most models it is impossible to plant in the sides. One type, not always easy to come by, does have holes in the sides.

The important considerations when it comes to compost are maximum water-holding capacity and light weight. Accordingly, peat-based potting composts are ideal, as the weight of soil-based types is liable to pull screws out of old brickwork. It is almost inevitable that they will dry out at some stage so drainage material to help re-wetting is useful. Vermiculite is the best solution as it has the helpful capacity to hold water itself as well as to allow good drainage. Branded composts specifically intended for hanging baskets are soon to be introduced in the UK.

Watering is crucial and if there is any doubt about your ability to water regularly stick to ivy-leaved geraniums which will put up with the lack of water. Baskets hung on warm walls are especially susceptible to drying out as the reflected heat from the wall magnifies the problem. A semi-automatic system such as that described for the tubs can be rigged up by pinning the tubing to the wall but it is a lot of bother. Moss-lined baskets can be removed and dunked in a large bath of water, but you would need a week or two in the gym first, so the simplest method is to use a hosepipe. Tie a 4ft (1.2m) cane to the end of the hosepipe leaving a 9in (23cm) overlap of pipe and you can easily hold the end of the pipe over the basket. Lances with angled heads are also available and these too make watering above head height easier.

Feeding is vital, with so little compost, and of course feeding in the usual way with a liquid feed in a watering can necessitates much precarious perching on step ladders. Feed units that fit in the hosepipe are, therefore, useful as are feedsticks which are put in the compost when the basket is planted and which steadily release plant food through the season. Slow-release fertilisers are becoming more widely available and are the most practical solutions.

When it comes to planting, trailing plants are obviously the first consideration but a dwarf, bushy plant in the middle of the basket together with a slow growing climber to train up the wires should also be considered. Baskets of a single plant in mixed colours work very well as long as the plants can go in the sides as well as the edge to create a real ball of colour. Petunias, impatiens, dianthus and fibrous begonias are especially suitable. Baskets in single colours are also excellent and here again impatiens, petunias and dianthus are good as are calceolarias.

| Good varieties to use in mixed colours are: | *Begonia* 'Stara Mixed' | **Summer** |
| | Geranium 'Ivy-leaved | **Baskets** |
| *Impatiens* 'Novette Mixed' | Mixed Colours' (from | |
| Mimulus 'Calypso Mixed' | cuttings) | |
| (regular watering is vital) | *Petunia* 'Picotee Mixed' | |

Good single colours
include:

Calceolaria 'Sunshine'
Impatiens 'Blitz'
(orange)
Dianthus 'Snowflake'
(fades early) (red and
white)

Begonia 'Pink Avalanche'
Coleus 'Scarlet Poncho'
Petunia 'Red Picotee' and
'Orange Bells'

Good combinations are:

Geranium 'Hollywood
Star'
Cerastium tomentosum
(silver foliage)
Lobelia 'Light Blue
Basket'

Geranium 'Video Mixed'
Campanula 'Krystal
Mixed'

Helichrysum petiolatum
(silver foliage)
Petunia 'Birthday
Celebration' (pink)

Begonia 'Pink Avalanche'
Impatiens 'Super Elfin
White'

Marigold 'Cinnabar' (rust)
or 'Solar Sulphur'
Parsley 'Bravour'
Pyrethrum 'Golden
Fleece'

Begonia 'Lucia' (mixed)
Campanula 'Krystal
Mixed'

Cerastium tomentosum
Campanula 'Krystal
Mixed'
Pyrethrum 'Golden Moss'

Petunia 'Recoverer White'
Helichrysum petiolatum
Lobelia 'Blue Cascade'

**Spring Baskets** In the last few years there has been a growth of interest in planting hanging baskets for spring display. Rock plants and bulbs are used, together with conventional spring bedders. The baskets are planted up in the autumn, over-wintered in a cold greenhouse and moved outside in spring. Good combinations include

Aubrieta 'Red Cascade' and Muscari 'Blue Spike'

Alyssum montanum (yellow) with Muscari 'Blue Spike'

*Viola* 'Universal Mixed'

*Cerastium tomentosum* with *Arenaria montana* (white)

*Arenaria montana* with *Viola* 'Universal Blue' or any red or blue pansy

*Arabis* 'Snow Peak' with *Armeria* 'Splendens' (pink)

## Window Boxes

While hanging baskets have become more popular, window boxes are seen less and less. This is largely due to the decline in the number of sash windows — plants obviously get in the way of windows which open outwards. However the country is still full of modernised Victorian houses the sash windows of which are intact or restored and so are ripe for boxes. But there is a second problem — plants in window boxes lean towards the sun and away from the house, so do not expect to get much of a show as you look out of the window — all you will see is a collection of bare stems. Be that as it may, when seen from the outside window boxes can be quite enchanting and well worth putting up.

The most vital thing of all is strong brackets. There is a lot of weight in a box full of wet compost and plants so it pays to be very thorough about the fixing. Generally speaking plastic boxes are a little on the small side, holding rather too little compost to support the number of plants grown. Homemade timber boxes are by far the most satisfactory and can then be constructed of sufficient depth and width to carry enough compost; 9in (23cm) timber is the minimum width. If your windowsills are sufficiently broad, the box can be rested on the sill, but you will probably find that the sill slopes to throw off water and to keep the box level blocks should be set under the front.

Drainage holes are vital and some plastic boxes are very meagrely supplied with them and some are without them altogether — either way they must be drilled. If you want to crack the plastic, use a punch. Drainage holes in timber boxes should be made with a brace and bit and all treatment with preservative done after cutting and drilling but before fitting together. A shallow layer of drainage material is essential, if only to prevent the compost falling through the drainage holes, but if you are worried about reducing the amount of compost, a piece of very fine mesh plastic netting, such as is used for greenhouse shading, will keep the compost in. In larger boxes a layer of expanded clay granules or coarse perlite will give drainage without weight.

*Figure 6.3 In deep window boxes, a layer of gravel will help surplus water drain away effectively*

When it comes to compost the same considerations as suggested for baskets should be borne in mind as weight and water retention are obviously important factors. The same plants and combinations recommended for baskets are suitable but one useful trick with boxes, which can be adapted for baskets with some planning and constructions, is to have more boxes than you have brackets to hold them. That way, by standing boxes not in use in a sunny spot, you can move the box looking the best into position while the others are prepared for a show a week or two later. It will also be possible to start off plants for the next season while the current season's display is still going strong and this will help make the best of each position. A box or two can be moved into the greenhouse or conservatory too and this will bring them on sufficiently for you to put them in place with plenty of colour in evidence. Feeding and watering is much the same as for the baskets and if you are clever you can rig up some semi-automatic watering for boxes too. Ideas suggested for tubs and hanging baskets are very suitable for window boxes.

# CHAPTER 7
# A–Z Annuals and Bedding Plants

## A

**Abutilon**

Two types of abutilon are grown for summer display. Both are tender but only one can be raised afresh from seed each year while the other is grown from cuttings.

The 'Bella Series' are perennial shrubs in some areas but are intended to be raised from seed each year. Sow early at 70°F (21°C) and plant out from 3in (7.5cm) pots. The flowers are like large bells and come in various reds, oranges, yellows and whites. They are held out slightly sideways to accentuate the display. In warmer climates they make a bush 3–4ft (0.90–1.05m) high but in an average British summer only reach a 1½ft (45cm) maximum. They are probably best in containers in a sheltered spot, but really come into their own as a conservatory plant where they will make good sized, branched bushes covered in flowers.

*A. striatum* 'Thompsonii' is an unusual variegated plant which needs protection from frost in the winter and must be raised from cuttings.

**Ageratum (Floss Flower)**

Although now mostly used in bedding displays, ageratum was once very popular as a cut flower and there are still one or two suitable varieties available. The colour range amongst the bedding types is slowly expanding from blue and dirty white to include some pinkish shades, lilac, purple and ice blue. Unfortunately, as the flowers die they turn brown and stay on the plants and no one has the patience to remove so many. This does make the whites look especially tatty. Beware of the illustrations of ageratums in catalogues — theirs is a blue shade singularly unsuited to colour reproduction and pictures can be very misleading. Some catalogues do not illustrate them at all, others try to correct the colour. Either way, they are not very helpful.

Amongst the bedding types, a number of F1 hybrid varieties from Holland have recently appeared and are named after various oceans. Although 'Baltic' and 'Adriatic' sound as good as they are, 'North Sea' conjures up a murkier idea of the colour than its deepest blue really justifies. Needless to say the mixture is called 'Seven Seas'. 'Atlantic' is mid-blue and flowers early, 'Adriatic' is slightly paler and at 6in (15cm) rather shorter than 'Atlantic'. 'Ocean' is pale blue and especially early to flower while 'North Sea' is the darkest, almost purple in colour, but is often rather too small and dumpy. The flowers are slightly paler than the dark buds. 'Ocean' was Highly Commended in RHS trials in 1981 while the open pollinated variety 'Blue Mink' received an Award of Merit. This is still probably the favourite variety, especially as it costs so little compared with the F1 hybrids. It tends to grow rather taller than the catalogues will tell you, up to 12in (30cm), but flowers wonderfully. It differs from the F1s also in that it does not make quite such an even plant — but that is hardly a disaster when you are only paying 10 per cent of the F1 price.

As far as white varieties are concerned they are justifiably ignored by most people. The problem is that the flowers go a rather pale but dirty brown as they die. Of course the same thing happens to the blue varieties but it is by contrast with the white flowers that it really shows up. 'Summer Snow' is about the only one anybody bothers to grow but almost any other white-flowered plant is an improvement.

Other varieties worth looking out for include 'Blue Danube' an American variety which was Highly Commended in RHS trials. The individual flowers are slightly smaller than those of other varieties but this is more than compensated for by the vast numbers in which they appear. 'Bengali' is an unusual lilac pink which becomes more and more pinkish as the flowers age. Finally 'Blue Ribbon' has recently appeared on the market, and this is especially distinguished by its very rounded habit with the flowers being produced down the sides of the plants as well as across the top.

The types suitable for cutting reach 18in (45cm) or sometimes more and are well branched with stiff stems. 'Wonder' and 'White Wonder' are about the only ones available though you may sometimes see the original wild species from Mexico, *A. houstonianum*, in some catalogues and this is rather more dainty. Both these varieties, but especially the blue, are well worth growing in borders too as blue summer flowers are in short supply.

The shade of blue in ageratums is never as intense as, say, lobelias and so they look well with pastel arrangements involving petunias, silver foliage of various sorts, white alyssum and yellow-foliaged pyrethrum. The taller varieties associate well with agrostemma and gypsophila.

The usual half-hardy treatment serves them well but leave the seed uncovered if you can ensure even moisture.

## Agrostemma (Corn Cockle)

Agrostemma is a native of the Mediterranean, which was once common as a cornfield weed in Britain along with the corn marigold (*Chrysanthemum segetum*) and cornflower (*Centaurea cyanus*). It is now much reduced by modern agricultural practice, although it can still occasionally be found in some parts of Britain.

There are three forms to be found in cultivation these days. There is the more or less wild form which is available from specialists in wild flower seeds and there is a cultivated version 'Milas', so called because it was originally found near the town of Milas in Turkey. The normal species grows up to 3½ft (1.06cm) and is very upright in growth with slightly downy stems and long narrow foliage. The flowers are five-petalled and form a loose trumpet in a soft reddish purple. 'Milas' is more of a plummy shade and altogether shorter at about 2½ft (75cm) with the colour grading to white in the centre. Some strains are especially stocky. If you can find it, try 'Milas Cerise' which describes itself.

An excellent cut flower and lovely in loose cottagey borders, its tall spindly growth makes it ideal for peeping through shrubs as it is rather straggly to grow in a group. Dead-head for most of the summer but leave the last few pods to scatter seed for the next year. Take care of the seedlings when hoeing through the border. The seeds are very short-lived and if you lose the first crop of seedlings, you may have lost the lot.

Corn cockle prefers a soil that is not too heavy but has managed to adapt itself to a variety of soils as it has spread through the cereal growing areas of the temperate regions. As long as it gets plenty of sunshine, it should thrive.

**Althaea**
**(Hollyhock)**

Hollyhocks are classic cottage garden flowers. Tall and stately, they were once commonly seen planted in mixed cottage borders and with climbing roses by the door. Sadly, they are less common than they once were and one of the main reasons, which has also caused the decline in popularity of antirrhinums, is the prevalence of rust. Unfortunately, there is not even partial resistance in hollyhocks and so fewer and fewer are grown.

Rust shows as brown pustules on the stems and leaves and spores spread from these pustules in large quantities in spring, infecting healthy plants very quickly. The spores overwinter on the leaves and leaf stems close to the ground so it is no bad thing to cut all these off in the autumn, collect up any other leaves and give a drench of propicon-azade fungicide to the crowns. In spring, spray the new foliage every couple of weeks. This is very effective against most rusts but you may have to spray regularly right through the season because if there are unsprayed plants in neighbouring gardens yours can easily be re-infected.

When mixing up the spray it helps to add a few drops of washing up liquid as a wetting agent. Hollyhock leaves are hairy and sprays do not usually spread very well unless a wetting agent is added. And do not forget to spray the undersides and the stems too.

Bear in mind that although they can live for a few years hollyhocks are best grown as annuals or biennials and it pays to dig up and burn plants after flowering.

The decline of hollyhock popularity has led to much less seed being sold, and unfortunately the quality of some strains has declined too. This is true of both singles and doubles and now not all the seed companies even list singles. This is a pity as it was the singles that were such traditional favourites. And it is in the doubles that there is such a variation in quality. But one development which has to some extent halted the decline in popularity is the introduction of dwarf varieties intended to be grown as half-hardy annuals.

There are just two dwarf varieties which are excellent plants for windy situations and smaller gardens. They are also strong growers and so can survive mild rust attack without too much trouble. 'Majorette' was the first to be introduced and is also the most dwarf. It grows to just 2½ft (75cm) at most and the double flowers can be as much as 4½in (11cm) across. They come in a good range of bright colours — various pinks, carmine, yellow, white, apricot and cream. They are best sown in mid-February at about 60°F (15°C) and it is important that they be kept growing well and not checked. Sowing two seeds in a 3in (7.5cm) pot is therefore a good idea, one seedling being removed from each. Do not always remove the weakest as this may influence the range of colours you end up with. Harden them off well, plant out in May and they should flower from mid-July onwards. They prefer a sunny site with fairly well-drained soil but avoid soil which is too rich otherwise the foliage has a tendency to grow too lush and hide the flowers. Make sure to remove the plants in the autumn so that rust is not overwintered.

The latest variety, 'Pinafore', is rather taller at around 3–3½ft (0.9–1.05m) but has the very useful quality of bushing out quickly from the base and throwing up five or six flower stems. The flowers are semi-double or sometimes single, very lacy and they come in a range of soft pastel shades — and there are plenty of flowers on a stem too. They are raised and grown in the same way as 'Majorette'.

To grow hollyhocks as biennials, sow in May or June in a spare part of the vegetable plot or at the back of the border. Make sure you sow thinly and then thin out to about 9in (23cm). Plant them in September. It has also been suggested that to get really strong plants, they should be lifted in October, potted into 5in (12.5cm) pots and overwintered in a cold greenhouse. They are then planted out in April where they rapidly make very good plants. They must be sprayed regularly during the winter months to keep rust at bay.

Amongst the more traditional types there are a number of single- and double-flowered varieties. 'Chaters Double' makes tall plants, up to 6ft (1.8m) with mostly double flowers in peony form and in strong bright colours. It is usually available in a mixture but you can sometimes find single colours. 'Summer Carnival' is about as tall with double flowers and can be treated as an annual or biennial. 'Single Mixed' makes the tallest plants, 6–8ft (1.8–4m), with a very wide range of colours and large single flowers; 'Powderpuffs' are also tall with the densely double flowers closely packed on the stems. Finally another small type, but one which is very rarely found. 'Silver Puffs' is very dwarf, 2ft (60cm), with vast quantities of delicate, silvery pink double flowers.

In the garden the tall types can best be used as part of the 'cottage tumble'. They will often need tying up to a wall or fence to keep them from collapsing into the border but if this should happen, you sometimes find new flower shoots grow up from the old stem making a row of short flowering spikes along the original stem.

**Alyssum**
(*Lobularia
maritima*)

Another plant from the Mediterranean which has established itself along the south-west coast of England, this sweetly scented little spreader is still usually known by its old botanical name of alyssum. The fact that its scent is sweet is sometimes overlooked, though its old English name is sweet alison — an appellation which combines recognition of its perfume with a corruption of its old scientific name.

In cultivation this is one of the best edging plants and an ideal one for small gaps at the front of borders. It can be sown outside from April to June, or indeed in September to overwinter. It will often sow itself in the autumn although, whichever variety you start with, the succeeding generations are likely to get progressively nearer to its height in the wild — 12in (30cm) — and nearer to its natural white colour. It needs sunshine and a soil that is not too heavy, though it will grow, in a rather straggly fashion, in shade.

Although basically a delightful little plant in most ways — the seed is even fairly cheap — it does have its problems. One is mildew, especially in hot summers. The other problem, also especially noticeable in hot summers, is that it can burn out early in the season, especially if sown early and even more especially if grown in a hanging basket. Here it gives a good show early on but once the geranium roots fill the compost and suck out all the water it rapidly crisps to a brown cinder. This is not so very bad in a basket as there will be petunias and/or geraniums to mask it but a patch of scorched foliage in the border is not pretty. So put some more in an odd corner in June and transplant the seedlings or pop in something else fairly quick-growing like mimulus.

Looking around the trial grounds another problem is revealed. Seed of many of the older favourites like 'Carpet of Snow' and 'Snowdrift' is raised by a variety of growers in different parts of the world. Owing to the fact that they have different ideas of what the variety should actually be like, as well as varying standards of roguing to remove off types, the plants raised from seeds from different sources may vary significantly. Plants raised from seed from one source may be much more spreading, slightly taller and much less floriferous than the dense, flat, well-covered plants from another. I can only suggest that if you think your alyssum is poor one year, try again.

**Amaranthus (Love Lies Bleeding)**

The dangling tassels of *Amaranthus caudatus* always cause interest and it is still grown widely both in cottage-style borders and annual borders as well as for flower arranging. The standard red-tasselled variety grows to about 2ft (60m) and the long red tassels hang down almost to the ground in elegant millet-like tails. The amount of seed produced is vast and once you have it in the garden you will always have it. The green variety, 'Viridis', is the same height but the foliage is a little paler and the tassels bright green, fading to cream.

There are erect versions too which are becoming increasingly popular especially with flower arrangers. 'Pygmy Torch' reaches 12–15in (30–38cm) with upright crimson heads of flowers and purple leaves and 'Green Thumb' is similar except that the spikes are brilliant green. *A. paniculatus* is altogether taller, sometimes reaching 3–4ft (0.9–1.2m) and as well as the upright purple plumes it has red-purple foliage. It tends to grow as a single stem unless pinched out fairly early. 'Red Cathedral' and 'Red Fox' are named varieties sometimes seen. A variety called 'Cruentus' is also sometimes seen and this has the same dark foliage but with pendulous flowers.

All these are quite happy with the usual hardy annual treatment and vast carpets of self-sown seedlings will appear if the seed is allowed to be shed. All are good for drying but the pendulous types cannot be simply hung upside down as they will not set in a natural way. Instead, pick them when about half the flowers are open, remove all the leaves and stand them in bottles or jars with the tassels hanging down. You may need some gravel in the bottom to prevent the jars falling over.

*Figure 7.1*
Amaranthus
caudatus

All the pendulous types make very good container plants. You do not need an enormous tub, an 8in (20cm) pot will do, but put just one plant in each pot and pinch the top out. By mid-summer you will have a spreading plant with stiff, arching branches and long tassels hanging down well below the rim of the pot.

As well as these fairly tough, flowering types there is an increasing number of varieties of *A. tricolor* grown for the foliage. These are most definitely half-hardy annuals and what is more they need fairly sheltered conditions outside.

A hot site, protected from the worst of the winds and a soil that is reasonably rich and which never gets too dry is needed — almost impossible I know and so a container may be the best place to grow these varieties. 'Illumination' is probably the most frequently seen and this makes a plant about 1½ft (45cm) high with foliage in red, bronze and yellow. 'Joseph's Coat' is the most dramatic, the leaves are yellow and red but this one especially needs a good site. A number of others will be found in catalogues but in less favourable parts of Britain it will pay to grow them as cold greenhouse or conservatory pot plants. This group resents too much root disturbance so prick them out into pots from which they can go outside.

**Anagallis (Pimpernel)**

The pimpernel or poor man's weather glass (*Anagallis arvensis*) is a weed in gardens and arable fields where its briliant red flowers open in fine weather and close when it is dull. The flowers are only about ¼in (6mm) across but their intensity of colour and their endearing habit of reacting to the weather has always made them favourites. Known also as the scarlet pimpernel it has many curious country names including John-go-to-bed-at-noon (which is curious as noon is the time it is most likely to be open), shepherd's clock, wink and peep and cry baby crab. It is also supposed to have many special powers including the giving of second sight and holding it is supposed to give the power to understand the speech of birds and animals. Having just been out with a handful to have a chat with the cat I can only testify that she said 'Miaow'.

It is such a pretty weed and such a harmless one that I rarely pull it out but there are some more truly ornamental forms which earn their place rather more honestly. These are usually varieties of *Anagallis linifolia*, a perennial from the Mediterranean which rarely overwinters outside in Britain so is usually grown as an annual. In warmer countries it will survive happily. In the improved form which is grown in Britain, the flowers are about 1in (2.5cm) across and brilliant gentian blue but there is also a mixture available with a range of shades, mostly pinks, blues and lilacs. Well-drained soils in sunny places seem to suit them best and this makes them one of the small group of annuals that really do look at home on the rock garden. Forget the Virginia stocks and the alyssum, the more delicate habit of anagallis is far more appropriate.

Seed should be sown in spring and treated as a half-hardy annual. The seedlings are especially small and delicate so they need careful treatment when pricking out. Standard peat-based composts can be a little too wet so add some perlite or sharp sand at about one part to three of the proprietary compost. Plant after the last frost date in the spaces amongst alpines, in dryish troughs and in the edges of gravel paths.

**Anchusa (Bugloss)**

As distinct from viper's bugloss (echium), which is stiffly hairy, anchusa has a smooth skin but still comes in some variety. There is a splendid (4½ft) 1.3m high perennial, 'Dropmore', which is a fine border plant but this is a variety of the southern European *A. azurea.* The ones normally grown as annuals are varieties of *A. capensis* from South Africa, a biennial in the wild and indeed its successors can be grown as biennials for spring flowering. More usually, though, they are grown as annuals, raised in a little warmth in spring and planted out around the last frost date or indeed sown where they are to flower.

The great thing about them is their colour — the most intense blue. And of course there are not that many good blue summer flowers. The two varieties normally seen are 'Blue Bird' and 'Blue Angel'. 'Blue Angel' reaches about 9in (23cm) and is rather upright in growth, making brilliant ultramarine plants which are densely compact and well covered in flower. Unfortunately the foliage is rather poor. 'Blue Bird' is more of a 'new denim' shade and the plants will reach 18in (45cm).

If you like the old red, white and blue colour schemes then bring these in, particularly 'Blue Bird' which at least enables you to get away from the red salvias fronted by alternate blue lobelia and white alyssum. Put the anchusa at the back and then scratch your head for something dwarf and scarlet. The shorter 'Blue Angel' also makes a useful cold greenhouse pot plant for autumn or early spring. Sow in summer or autumn.

**Anoda**

Anodas are rarely seen, for no other reason, it seems, than they are not terribly trendy. From Central America, they can be thought of as annual versions of hardy geraniums. They are quite hardy and can be sown outside in spring as hardy annuals although in colder areas it would be wise to start them off under a little protection. There is only one species normally grown and unlike some of the rarer ones it prefers soil that is not too wet and not overfertile. In wet summers flowering may be depressed or delayed and in areas where the summers are not fine, it may pay to sow in the autumn and overwinter in a cold frame or greenhouse. This may seem overfussy attention for a relatively unknown plant, but it is worth it. The plant grows to about 2½ft (75cm) and is very upright in growth. The flowers of the variety usually seen, 'Opal Cup', are 2in (5cm) across, silvery lilac in colour with slightly darker veins and they are altogether very delicate. It is a good plant for the mixed border and at first sight seems an ideal companion for agrostemma but some people object to the association of the two colours.

## Antirrhinums (Snapdragon)

Over the last 20 years, fewer and fewer people have been growing antirrhinums even though, in theory, they are such splendid plants. There is only one reason for it — rust. This is a fungus disease which first appeared in Britain in 1933 and which quickly spread to the whole country. Fortunately there were some rust-resistant varieties of antirrhinum grown in America, where the disease originated, and these were brought over to Britain. However, five years after the arrival of the disease only three varieties maintained their resistance. Then, not long after, a new strain of the disease arrived and even these three resistant ones succumbed.

Advances in breeding rust-resistant varieties were made at the Royal Horticultural Society at Wisley and then at Hursts by Ralph Gould, still one of the country's foremost plant breeders. In the late 1950s and early 1960s he introduced a range of varieties all with the suffix 'Monarch' and some of these are still amongst the most resistant available. Occasional varieties have been raised since then, especially by Suttons, but new strains of the disease keep arising and the resistance breaks down. None of the varieties available now is totally resistant, all that can be said is that if you choose the right ones they will be affected by the disease so little that a good display is possible.

The most recent investigation of the disease was in a joint study at Wisley by the RHS and Royal Holloway College, London and they recommended 16 varieties out of the hundreds they grew. Since the trial in the late 1970s some have disappeared and the appearance of new strains of the disease threatens to cause more problems. New varieties of antirrhinum have also been introduced but none seems to be resistant. On the strength of the suggestions that came out of the research the following readily available varieties can be recommended. Although there are only six, they cover most of the colours available in antirrhinums.

'White Monarch' — Pure white flowers on plants 15–18in (38–45cm) high.
'Suttons Rust Resistant Yellow' — Bright yellow flowers on plants 15in (38cm) high.
'Orange Glow' — Orange with a pink throat and growing to around 15in (38cm).
'Leonard Sutton' — Rose pink flowers starting earlier than most on plants 15in (38cm) high.
'Scarlet Monarch' — Brilliant scarlet flowers on plants around 15in (38cm) tall.
'Crimson Monarch' — Dark crimson flowers on plants 15–18in (38–45cm) high.

The important lesson to learn from this is that mixtures are dangerous. Supposedly rust-resistant mixtures, like 'Coronette Mixed', contain some colours which are relatively resistant and some with very poor resistance. The result is that the poor ones quickly become infected and their neighbours are then overwhelmed by the volume of infection around them. Unfortunately, the separate colours are rarely available, but the white, bronze and cherry are the most reliable.

This leads to another point. If you leave plants in through the winter the disease will overwinter on them and provide a strong source of infection in spring when new plants are put out. So it is wise to burn all antirrhinums in the autumn whether they seem to be infected or not. The rust problem has been somewhat alleviated by the recent introduction on the amateur market of propiconazole, a chemical which will kill rust. It is the most effective fungicide for rusts on ornamental plants but it must be sprayed on the plants early. There is no need to spray as a precaution but it is wise to learn to recognise the first signs and spray them.

The first symptoms are small, pale spots on the undersides of the leaves. These become redder as the disease takes a hold and eventually erupt into brown pustules which release spores into the air. These spores then infect other plants. At this stage pale patches appear on the upper surfaces of the leaves. Both leaves and stems can be affected. In severe cases leaves and stems eventually go brown and the leaves drop off leaving an ugly, twiggy plant. The time to spray is when the first signs appear and excellent control can be achieved, but there is a small snag. You need to spray at the first sign and then again three more times at fortnightly intervals. If this regime is followed your spray will do an excellent job. Research shows that sprayed plants had their infection reduced from just over 1½ per cent to less than ½ per cent, which is negligible, while on unsprayed plants the infection rose from just over ½ per cent to 75 per cent! But it is vital to keep up the sprays.

There is more good news on the horizon. Following the research at Wisley, new plant breeding work is under way and it may not be long before we have a whole new generation of rust-resistant varieties.

But that is not the end of the difficulties you are faced with when growing antirrhinums. In spite of being favourites with gardeners since the last century, many people still have difficulty raising them. It pays to sow on the surface and leave them uncovered, but the surface must not be allowed to dry out. Damping off is usually the problem and antirrhinums are especially susceptible. Dealing with damping off is described on page 39.

There are a lot of precautions you must take and this seems strange, especially when you recall that you can sow antirrhinums outside in the open ground and transplant them to their flowering sites or thin them out for flowering *in situ*. Of course the humidity and high temperatures which cause such problems in the greenhouse are not present outside but you will still find that a large proportion of the seed you sow will not appear. Fortunately sowing outside is much less common than it was, especially since autumn sowing is so frowned upon because of the problem of overwintering rust.

Continuing the litany of complaints against this splendid plant we come to the flower problem. Most modern varieties will produce a good spike of flowers, in some cases a large number of good spikes, soon after planting out. The problem comes next: that is often the end of the display for quite some time, especially in hot summers. The problem is worse in municipal bedding and in parks where there is not the time to dead-head the plants but even at home you will usually find that there is a significant gap in the display between the first flush and successive bursts. But it is vital to remove the dead flower spike the moment the last flower fades, if not before, otherwise disappointment is guaranteed. Of all the plants described in this book this is the one where dead-heading makes the most difference. Often, as the flowers on the first spikes fade, you can see the tiny buds that will produce the next flowering shoots low down on the plant. If the seed capsules are left to develop, these young buds only develop very slowly and finally, when they open, there will only be a few flowers, often in rather a loose spike. Cut the first heads off promptly and the next ones will develop much more rapidly leading to more flowers in the head.

That is about the end of the problems, now let us come on to the good things — and fortunately there are quite a few. First, antirrhinums will thrive in a wide variety of soils. As long as it is not too acid, any soil will suit them though heavy clays should be improved to make life easier for their intensely fibrous root systems. They are best in sun but will do fairly well in semi-shade and the dwarf varieties will even produce a creditable, though rather lax, show in full but open shade.

Of course the dramatic variety in the flower colour, even without a true blue, and the captivating form of the flowers are all important — from which you will gather that I am not a great fan of the open-flowered or so-called azalea-flowered types. The colours are intense with glorious bicolours or they may be delicate and soft. The carpeters are covered with masses of short spikes so you can hardly see the leaves and the taller types have long spikes with the flowers elbowing each other aside, there are so many. Apart from the specially rust-resistant types listed earlier there is an enormous range growing from 3ft (90cm) to 6in (15cm). Having grown a good few and seen most I would recommend the following.

Amongst the tall growing types for cutting or for the centre of the large beds there are three main varieties. First of all there is 'Madame Butterfly'. Growing to about 2½ft (75cm), the flowers are like double azaleas and about 2in (5cm) across. The colour range is very wide and includes both strong and pastel shades. In the more conventional snapdragon-flowered type there are two varieties usually found, 'Giant Forerunner' and 'Spring Giant'. They are very similar and both grow to 3ft (90cm) or just over and branch well from low down giving long stems for cutting. Remember when cutting these tall cut flower types that if you leave them lying on the ground or worktop for too long or even lay them in a bowl at an angle, the tops will curve towards the light and you might loose the stately elegance of the spikes.

Most of the best garden varieties fall into the intermediate height range growing to about 18in (45cm). For some years most of the development in antirrhinums was geared towards producing F1 hybrids and indeed the best two varieties fall into this category. But recently less expensive types have been developed which are almost as good, though the 'Minaret' series still stands out. The first flush is the earliest of all varieties and, unlike many, consists not just of one central spike but of a number all round the plant. By the time they are past their best, the flowers on other varieties are approaching their peak. Later even more spikes develop so that I have had eight spikes on the first flush and late in the season as many as 24 open on one plant. The colour range consists of dark red, orange scarlet, orange, yellow, purple, lavender with a white centre, rose pink and white; but it is unusual to find the colours available separately, you will usually be stuck with the mixture.

The other good F1 variety is the 'Coronette' series. This behaves in a rather different way and produces one central spike which is fairly impressive but usually sits in the middle of a clump of green shoots. When the first spike is fading and removed, the show really gets going and in August they are usually more impressive than the 'Minarets'. There are nine colours; apricot bronze, scarlet, cherry, crimson, pink, rose, ruby, yellow and white. Some of these colours are occasionally available separately but again it is the mixtures you will most often find. The spikes are especially elegant and although only reaching 16–18 in (40–45cm) still make good cut flowers. With both these varieties it helps to get them out into the ground before the rest of the bedding. They will stand a little frost and will flower much sooner.

Two recent introductions in non-F1 types are 'Cinderella' and 'Vanity Fair'. I have grown 'Vanity Fair', which is an excellent mixture in a wide range of colours including bicolours. It grows to about 18in (45cm) and although I did not find it ever had quite the impact of some of the F1s, neither did it have quite such dull periods without flower.

'Cinderella' is about the same height, has about half-a-dozen strong colours and, like the 'Coronettes', tends to produce one spike in the centre followed by a mass of new spikes a little later. The great virtue of these two varieties, of course, is that they are a lot cheaper and so splendid value for money.

In addition to those recommended as especially rust-resistant, which all fall into this intermediate group, there are four other single coloured varieties which I like, although the first two seem to be especially susceptible to rust I am afraid. 'Black Prince' has very dark maroon foliage and dusky red flowers and is lovely with other hot-coloured flowers or with white for a real contrast. 'White Wonder' is a variety you might consider as a neighbour, or I have seen the two mixed together quite effectively. The flowers are pure white and the foliage pale green. 'Rembrandt' is dark orange with yellow petal tips and is very striking but tends to flower rather late; finally 'Purple King', which is an interesting lilac-purple shade with an orange throat.

In the dwarf varieties, almost all the varieties available are mixtures. In spite of the fact that 'Orange Pixie' was one of the first Fleuroselect medal winners, even this can only be had in a mixture — and some 'Pixie' mixtures do not even contain the orange. Very odd. The 'Pixies' reach about 10in (25cm) and are of the more upward facing open-flowered type. They are very floriferous and the colour range is bright with two reds, orange, yellow, rosy pink and white. They are especially impressive when seen from a distance as their open flowers show off the colour more effectively. But they are less elegant than, for example, 'Floral Carpet' although flowering starts much earlier. 'Floral Carpet' reaches about the same height, maybe a little taller, with seven colours, but keeps up the flower power better than most and overall is probably the best of all the dwarfer types. The smallest available, at little more than 6in (15cm) is 'Little Gem' and the flowers are small as well. This variety does suffer from the old problem of gaps between flushes but a quick trim with the shears will help, plus water in dry spells. 'Trumpet Serenade' is an open-faced rival to 'Pixie' at a much lower cost, but looking around the trials there does seem to be some variability in the quality of the various stocks on offer. You will get some snapdragon-types and it can be very late to start flowering.

The most interesting of the few single colours available in the dwarf group is 'Delice'. This makes spreading, rather flat and then eventually more rounded plants up to 9in (23cm) high and the flowers are of the most unusual creamy apricot shade. The flower spikes are quite short but they cover the plants.

Finally, a plant to be warned about. There has recently re-appeared in one or two catalogues a variegated plant called *Antirrhinum majus* 'Variegata' which has white flowers. It makes a sickly looking plant and apart from its curiosity value there is not a great deal to be said for it. It must be propagated from cuttings and as it is not entirely hardy, cuttings must be overwintered in the greenhouse. This of course, means that there is a definite possibility of rust overwintering. I will not say do not grow it, but be warned.

## Arctotis (African Daisy)

One of a group of sun-loving annuals in the daisy family which includes dimorphotheca and venidium, all of which came from South Africa. They are united in their big bright daisy flowers, their preference for sun and well-drained soil and the vigour with which they grow when happy. Arctotis used to suffer from the irritating habit, as far as the gardener is concerned, of the flower folding to a pointed bud-like shape in late afternoon to protect pollen from heavy South African dews. The species still behave this way but some of the selected strains are rather more amenable. They make dramatic cut flowers but do not last as well as, say, gazanias. Give them the usual half-hardy annual treatment and they can also be sown outside direct. They often seem to do better pricked out into individual pots rather than boxes, as they do not like too much root disturbance.

There are some mixed hybrids to be had from seed companies and these will give you a brilliant range of colours including purple, blue, red, orange, yellow and creamy shades plus some in between. They will reach about 18in (45cm). You might come across *A. acaulis*, rather smaller at just 12in (30cm) and originally brilliant yellow and this colour still crops up in the mixtures which appear. The undersides of the flowers in wild and cultivated types are rich purple while the main flower colours in catalogues vary from orange to rich red. The undersides of the leaves are covered in long white hairs. Try it as a pot plant in 4 or 5in (19–15cm) pots too.

*A. venusta* is pale blue and has the endearing habit of following the sun round during the day and then nipping back to the east during the night to be ready for the dawn. *A. grandis* is sometimes seen too and this has glistening white flowers with purple centres and dusky purple tints underneath. The flowers can be 3in (7.5cm) across and the leaves are downy white. A splendid plant for big clumps towards the back of the border as it happily reaches at least 2ft (60cm).

**Argemone (Prickly Poppy)**

A large-flowered member of the poppy family with slightly sea green leaves. The leaves have a spine at their tips and there are often more spines on the other parts of the plant. They are mostly biennials in nature but usually grown as annuals in gardens. They are not entirely hardy but, because their fleshy roots tend to take exception to being transplanted, it pays to sow them outside where they are to flower — but a little later than hardies are normally sown, say towards the end of April.

Prickly poppies like very sunny, well-drained conditions — which is hardly surprising when you consider that they are native to the southern USA and Mexico. *A. grandiflora* is very dramatic — enormous white flowers up to 4in (10cm) across with the delicacy of most members of the poppy family. They do not last all that long but appear in such continuous profusion that the plant is never dull. It will reach almost 3ft (90cm) in a good year. *A. mexicana* is a little smaller in overall size and in flower size. The flowers, though, are a very special, slightly lemony shade. This is sometimes known as the devil's fig though quite why I am not sure (doubtless some learned correspondent will enlighten me). People tell me that these have an odd scent but this is one of the flowers whose scent I cannot detect.

Using these plants can be a problem. If grown in an ordinary fertile and well-watered border they will not thrive and flowering will not be dramatic. But few gardens have the hot, well-drained corners free for annuals. Use them as temporary replacements for the larger rock garden shrubs after fierce winters, and in new gardens they can be especially useful. In those areas where you have not had time to dig in masses of muck grow prickly poppies with Californian poppies and other adamant sun lovers.

**Asperula (Woodruff)**

In the great quest for annuals that do well in the shade the woodruff seems to have been unjustly neglected. There is

84

only the one annual species, *A. setosa azurea* (also known as *A. orientalis*) but it is a little treasure. It grows to about 12in (30cm) and has dense puffs of pale powder blue flowers with the most delicious scent. It is a sprawling plant, but none the worse for that, and it thrives in moist shady places even under trees. Being quite hardy it can be sown where it is to flower and can be cut for the house too, lasting well in water. Its informality makes it a delightful plant for the mixed, cottage-style border, rather than for bedding and it is worth siting a clump near the patio too, for the scent.

## Aster (*Callistephus chinensis*)

There is really no rival to asters as late flowering bedding plants and they are now available in a very wide variety of forms from tall cut flowers to bushy little bedders. And when you consider that they are all developed from the species *Callistephus chinensis* introduced from China in 1731 you see that the improvement has been dramatic. The first one was a single-flowered purple but now a variety of petal and flower forms have arisen and most of these are fully double. But lovely as they are one problem has still not been entirely solved — wilt.

Aster wilt is a soil-borne disease which attacks asters after planting. It is a very specific problem and does not attack michaelmas daisies, or anything else for that matter — just asters. The fungus invades the roots and then travels through the plant up into the stems. As it goes, it blocks the tubes which transport water and food materials around the plant and at the same time the plant reacts by producing sticky gum in the tubes to try and stop the fungus spread. Unfortunately this gum also blocks the tubes, sap flow is halted and the plant wilts.

The plant may grow very slowly or grow in a rather one-sided way if the roots on only one side are attacked. One side of the plant may die entirely before the problem spreads to the other side. Sometimes the collapse is sudden and dramatic, especially in dry conditions, or if the attack is not severe, the plants may wilt during the day and recover by the following morning before wilting again. If you suspect wilt, cut open the stems above ground level and you will see a ring of brown stain where the fungus has blocked the sap channels. A whole bed of plants can be affected together if the soil is badly infected and the display can be a complete disaster.

Unfortunately, once you have the disease it will persist in the soil for many years. The fungus produces special spores which lie dormant in the soil until 'woken up' by root secretions from asters; they then attack the roots. It pays to wait many years, seven or eight at least, before replanting where wilt has been a problem. Wilt cannot be cured chemically and the only real answer is to wait and grow something else in the meantime. But you can guard against the problem with a number of sensible precautions. Only raise asters in proprietary composts, do not use a mixture containing soil from your garden, or worse, from anybody else's; do not plant asters raised from seed grown outside in a neighbour's garden and if you know a friend or neighbour's garden suffers from wilt don't walk all over their soil and then walk all over yours without washing your boots thoroughly.

Another useful trick if your soil is infected is to raise the plants in the normal way but instead of pricking them out into trays, use cellpacks or small pots. Then move them into 3½in (9cm) pots and then into 5in (12.5cm) pots before planting out mature plants. This is all rather arduous, especially as the pots will have to be moved outside at the end of May, as they would if planted in the normal way. This treatment will, of course, give you very strong plants and you will need slightly fewer than normal. Once planted the root ball of good sterile, wilt-free compost will sustain the plants for quite some time — at least long enough to produce a reasonable crop of flowers on even the most susceptible varieties. Needless to say, this procedure is only worthwhile for cut flower varieties, for bedding it's easier simply to choose a different plant. There are quite a few varieties which are said to be resistant but it is probable that none is 100 per cent reliable. This is compounded by the fact that some are only available through the wholesale trade to growers such as parks departments who may need to grow asters in the same place every year and for whom resistant varieties are especially useful. Different stocks of the same variety may also vary in their resistance. In wilt-free soil they can be fluid sown (see page 259).

There are two asters that are supposed to be more or less resistant. 'Ostrich Plume' is an old favourite and the name more or less describes the form of the flowers. These come in a good range of colours and although the plants are 18in (45cm) high, they are good bedders as well as useful for cutting. They are amongst the first to bloom and one of the best all-purpose types. 'Dwarf Queen' is also one of the earliest to flower, but a genuine bedder reaching about 10in (25cm). It makes very rounded plants with flowers well down the sides as well as over the top. The flowers come in red, purple, pink, lilac and white although the white is not always fully double. It is a lovely edger and good in containers too. It is also known as 'Carpet Ball'.

Going on to the less resistant bedders, 'Pinocchio' is even smaller, about 8in (23cm), and a Fleuroselect winner in 1974. It comes in red, purple, pink, lilac, yellow and white. No resistance to wilt I am afraid, but splendid in containers where they will be growing in new, sterile compost. You might come across 'Thousand Wonders', a very dwarf type, just 6in (15cm), in the usual colours; very neat and dainty with big flowers for the size of the plant. 'Teisa Stars' has double quilled flowers in nine colours on broad spreading plants, and there is the 'Milady' strain. This is very sturdy and vigorous in spite of its modest 12in (30cm) and so withstands wilt attack pretty well. The plants are upright with broad clusters of flowers in six colours — red, rose, a shade between the two, a pale pink, blue and white. The blue seems to be the earliest to flower.

Traditionally asters were always grown for cut flowers and they last so well that anyone with the smallest pretence to decorating the house with flowers ought to grow them specially. 'Duchess' and 'Princess' are the old favourites, doubles and singles respectively. 'Duchess' makes about 2ft (60cm). The petals are incurved like a chrysanthemum and they last very well in water. They come in deep red, crimson, two pinks, a dark and pale blue, yellow and white. These are as resistant to wilt as many, but strains vary. The plants are very upright which makes them ideal for cutting and in many gardens they will not need any support. Single colours are sometimes available. The single 'Princess' strain, which often appears in variants such as 'Super Sinensis' and 'Madeleine', is fairly resistant and at best a superb cut flower with bright yellow discs at the centre of flowers which come in just about every aster colour. The 'Rivieras' also make about 2ft (60cm) with slightly incurved petals on very large flowers. The stems are specially stiff and upright. The 'Brixham Belles', also known as 'Flamir', are taller at about 3ft (90cm) and come in a relatively limited range of colours with little in the blues and purples.

**Atriplex (Red Orache)**

There are not all that many purple-leaved plants that fit within our scope so even the less than perfect are worth making the most of. *A. hortensis* 'Rubra' is a relative of the spinach with bright, purplish red foliage and is an easy hardy annual, seeding itself about harmlessly, leading to some curious combinations. It grows to about 4ft (1.2m) and if not pinched will sometimes only produce the one single shoot in doing so. Once there were coppery varieties, pinkish varieties and one with darker veins but the beetroot variety, now the only one found, is good in flower arrangements but also with pink and white plants outside. In dry situations it rapidly loses its lower leaves and can look unreasonably spindly — but it is worth growing for all that. Try a few leaves in salads too.

**Aubrieta**

One of the most reliable and most common of rock garden plants which, like forsythia, is sometimes despised for being so commonly planted — which is only the result of aubretia's being such an attractive, easy and reliable plant. Never regularly used as a bedding plant, there is a good argument for treating it in this way. The colour is so intense and it is so often seen in other gardens that I usually like to use the space for something else. But I miss its intensity so every few years I use it in a bedding scheme with tulips, pansies or polyanthus.

Until recently, stock plants were needed as the seed-raised strains were rather watery in colour and variable too. Now there are two seed-raised varieties that really do deserve to be grown, 'Ruby Cascade' and 'Blue Cascade'. There is also a new F1 hybrid variety, 'Novalis Blue', which is constantly being listed in catalogues and which often proves to be unavailable as the seed crop has failed. It is a lovely lilac blue shade and the flowers are almost twice the size of most varieties. But when seed is available it costs ten times as much as older varieties. The two cascades, though, are the best bet at present.

The seed is sown in June in a cold or slightly heated greenhouse, pricked out into 3in (7.5cm) pots and stood outside or in a cold frame with the glass removed all summer. One of the problems of raising in pots is that it is easy to leave the plants in the greenhouse too long and straggly plants result. They do need to be grown hard. In the autumn they are planted about 6–9in (15–20cm) apart depending on how much they have grown. They like sun and a soil that is not too soggy in the winter. I grow the 'Ruby Cascade' with white muscari or strong yellow 'Tête à Tête' daffodils and 'Blue Cascade' with more of a primrose daffodil like 'Hawera' or with a tulip such as 'Franz Lehar'; polyanthus too are good companions. Aubrieta is one of the spring plants that look good in baskets. Plant up in the autumn with the same combinations, keep in a cold greenhouse until March and then hang them in position outside.

# B

**Begonia**

Considering the great number of varieties that are available from all over the world, the seed companies are commendably restrained in the range of begonias they offer in their catalogues. And with the varieties now available, it is difficult to be disappointed. Both the fibrous rooted types and the much more flamboyant tuberous rooted kinds, though neither cheap to buy nor the easiest plants to raise, make such reliable and such sparkling garden plants that the expense and effort is worthwhile. And when the tuberous variety 'Clips' first appeared the seed was more expensive than gold so you will know they are not cheap!

*Fibrous Rooted Types*

These were developed mostly from the Brazilian *B. semperflorens.* In the wild this makes a plant up to 18in (45cm) tall with green leaves and white or rose-tinted flowers. Much of the breeding effort has been geared to reducing the size of the plant and increasing the colour range and of course the development of dark-leaved varieties has added quite a new dimension. The accent was on prolific flowering even if the flowers were not big. As the overall size of the plants was reduced, so the flowers became smaller too although recently this trend has been reversed.

Perhaps the finest achievement so far in fibrous begonias are two series of varieties from Holland, the 'Coco' series and the 'Verdo' series. As you might guess the former has bronze foliage and the latter green. Both types grow about 6in (15cm) and carry masses of flowers. The 'Coco' series includes scarlet, three different pinks and a red and white bicolour while the 'Verdo' contains scarlet, salmon, three soft pinks and a white. The particular colours available to the home gardener vary according to the whims of the seed companies. As you might expect the two strains look very good mixed together and a number of companies list a mixture. But therein lies a problem. The raisers of the varieties call their mixture 'Lucia' and most excellent it is too. They use all the colours in the two strains plus one or two intermediates, but they do not mix them equally. They tried different proportions until they created the one which they felt had the widest appeal. Some of the seed companies that sell to home gardeners agree with their selection and sell the same mixture under the same name. But others do not agree and prefer to alter the mixture slightly according to their own ideas of what the gardener likes. So you will also find 'Devon Gems', 'New Generation' and 'Fantasy Mixture' which are made up of the same colours in different proportions. It seems to be mainly the proportion of dark-leaved types and the proportion of whites in the mixture that causes disagreement amongst the experts. But the fact is that all the mixtures make lovely sequin-like carpets of colour and although you will soon discover that I do not have a great fondness for mixtures in general, mixtures of begonias, especially fibrous rooted ones, are an exception.

Looking a little more at the 'Lucia' group and others of the smallest growing fibrous types, their other great appeal is their tolerance of a variety of garden conditions. As long as the soil is not very dry they are happy in sun or shade, indeed many people consider them the equal of impatiens as shade-loving bedding. In wet summers they are good too, putting on a splendid, sparkling display that will encourage you to venture out with the brolly in the most unappealing of conditions.

Another variety worth looking out for in this small size range is 'Organdy Mixed', which if anything is rather smaller but has a similarly broad range of flower colours. As to the foliage, there is a little confusion as some companies have a mixture which contains both foliage colours, others have a mixture in green only. Read the packet or the catalogue.

A new development in plants at this smallest end of the range is the 'Trek' series also known as the 'Athena' series. These are four varieties from France which are less than 6in (15cm) tall but with very large flowers. Colours are rose, salmon, scarlet and white and this could be the beginning of a new range of very dwarf plants with a dramatic colour impact.

In single colours, the ones that are the most useful are 'Coco Bright Scarlet' whose combination of red and bronze is just right and the two other 'Coco' shades available, pink and bicolour. There is also 'Viva' with white flowers and fresh green foliage which is also good.

Going up in size a little we have the old 'Danica' series of varieties at about 10in (25cm). 'Danica Red' was for a long time the best of bedding varieties with brick red flowers and dark bronze leaves and it is the one I always choose for growing with chlorophytums. Just the red and also a rosy pink are available but they flower early and they are particularly resistant to spells of bad weather. Much more recent are the 'Frilly Dillies' (sorry about the name) which at present are available in 'Frilly Red' and 'Frilly Pink' and which reach about 12in (30cm). The flowers are large and the petals have pretty wavy edges. Their growth is noticeably upright and they branch well from the base making especially sturdy plants. I have found them better in relatively sheltered places rather than at the full mercy of the elements. I have not tried them in baskets but they should be very attractive.

Much better in baskets is another relatively recent one, 'Pink Avalanche'. Again reaching 12in (30cm), it has a rather loose cascading habit and the flowers are especially long-lasting and continually prolific as they set no seed. There is a white form too, and a mixture, and this is the strain I would always choose for baskets.

Although usually raised from seed sown fairly early in the spring, maybe at the start of February, to give them the long growing season they need to make substantial plants to go out in flower at the end of May, the very small seed creates problems for the home gardener. The main difficulties lie in uneven sowing, or at least lack of confidence in being able to sow evenly, covering too deeply and keeping the compost too wet. The expense of some seed is also offputting. But for fibrous rooted types there are other ways of growing them.

Many mail order seed companies now sell trays of begonia seedlings although only mixtures are to be had. There are sometimes too many seedlings for one garden but by ordering petunias and pansies as well, two or three neighbours can get sufficient of each type. Recently begonias have also become available to the home gardener in plugs, that is small, wedge-shaped blocks of compost each holding one young plant. They are rather more advanced than the seedlings in trays, rather fewer come in a pack and later in the spring, so the gardener does not have to worry about providing high temperatures for so long.

*Tuberous Rooted Types*

These dramatic plants can be grown from seed, from tubers or bought as young plants. The prices are interesting. Taking the price of a packet of 'Non Stop' seed as a base, ten young plants in Jiffy 7 pots cost almost twice as much as the seed while ten tubers cost about 1½ times as much as a packet of seed. Considering that you do not get a vast amount of seed in a packet, I would go for the tubers every time even if the display is not quite so dramatic.

The 'Non Stop' strain is exactly that. The flowers are up to 2½–3in (6.5–7.5cm) across and the plants grow to about 12in (30cm). As the name implies one of their most special characteristics is the fact that they start to flower when they are as small as 3in (7.5cm) and carry on and on, right into the autumn and if you move their container indoors before the frosts they will flower for even longer. What is more, there are nine different colours in the range — scarlet, bright red, salmon, two distinct pinks, orange, apricot, yellow and white. Whichever way you look at it they are expensive for use as bedding in the garden but make wonderful plants for tubs and window boxes. Indeed a tub with nothing but five 'Non Stop' begonias will give you all the colour you could ask for all summer long.

If you do raise them from seed and have one or two to spare, pot them into 5in (7.5cm) pots as they make splendid presents. Unfortunately, the colours are not entirely uniform. The scarlet and bright red seem to have the most flowers, the yellow tends to make smaller plants, a number of singles creep into the whites and one of the pinks is rather tall and upright.

There are other newer varieties available in the same mould, in particular 'Clips' and 'Memory'. 'Clips' is altogether less appealing. For a start there are only four colours and the plants are also much smaller. They are said to flower a little earlier than the 'Non Stops' but I have not found them to be an improvement. They seem rather variable in size, the flowers are smaller, singles creep in, though the yellow is a particularly intense shade. They are also more expensive. 'Memory' has slightly larger flowers than 'Non Stop' but is less branching and although there are twelve colours the overall impact is not as dramatic. There are fewer singles. At the Liverpool International Garden Festival, large beds of these three varieties were planted out alongside each other and the most striking thing of all was the impact of the 'Non Stop' varieties. It was especially easy to pick out the separate colours in the 'Non Stops' and it was easy to see why they are so popular. 'Clips' with just four colours and too much emphasis on the reddish shades was far less impressive.

Other seed-raised tuberous types worth a look include 'Cora', with large quantities of big, single, scarlet flowers on plants reaching just 10in (25cm) and they really are stunning. 'Pendula Chanson Mixed' is also worth trying if you are a fan of hanging baskets. Long, trailing stems carry enormous numbers of semi-double flowers and although the flowers are not as big as they might be, with six shades and such productivity the effect is very impressive. One new strain which should soon be available is called 'Midnight Beauties'. The foliage is dark purple, the flowers double, semi-double and single in a very intense orange. There is quite a number of other strains around and all will produce a reasonable display but the tubers will be the most reliable. 'Sensation Pendula' is probably the best hanging basket begonia of all — the flowers are fully double, but in just four colours, on long rather arching stems.

Raising both these groups from seed is a tricky business for the home gardener. The compost is important. Although it is true that the compost must never dry out — when the seedlings are small even a little drought at the surface can be fatal — if it lies too wet, damping off can be rife. So for begonias I prefer to use a compost with some sand or vermiculite to give a little extra air. Some composts have this as part of their formulation; if not, you can add it at a rate of about one part sharp sand to three of compost. Even sowing can also be tricky. It is often said that adding a few pinches of fine, silver sand to the seeds before sowing helps by giving you a little more bulk to sow. But the seed and sand separate so easily that you can sometimes find that although you have sown the sand evenly the seed still ends up all together on one side of the pot. It is also sometimes suggested that the sand be spread on the compost before sowing so you can see where the seed falls but I still cannot see it. So this is what I do.

I prepare the seed pot in the normal way making especially sure that the surface is level and even, then open the packet by cutting the top off with scissors. Now pinch a crease in the middle of one cut edge. By squeezing the sides of the packet carefully towards each other the seed will run into the new central groove. Make sure the seed lies evenly along this groove and then, tapping the packet gently with the finger, move the packet over the compost surface. The trick is not to watch the compost surface in the hope of seeing how the seed is being distributed. Instead, watch the seed as it falls out of the packet and then as you steadily move the packet across the surface, if a few taps fail to dislodge the seed you can stop until seed is sown. I should say that I do not claim this to be a perfect system — it is simply less imperfect than others.

As to covering the seed, very little covering is needed. Two taps of the sieve is usually enough, and many suggest not covering at all but the surface of the compost must not dry out for even an hour. The correct compost should help to solve the watering problem and as long as the young seedlings are protected from strong sun you should be successful.

**Bellis (Daisy)**    Now if there was a plant that declined in popularity as violas and pansies crept into more and more gardens it was the humble daisy — although in its finest cultivated forms it is not in the least humble. Daisies have come a long way since they made the short trip from the lawn to the border — the problem now is they do not hesitate to make the return trip, but more of that later.

Time was when bellis were amongst the most popular of spring bedding plants and there were a number of named varieties which were propagated by division each year: 'Alice', 'Dresden China' and 'Victoria' were some of the named forms. Then the seed-raised strains became rather more reliable particularly in terms of flower colour and form. They were also once grown as summer annuals, winter pot plants and even as cut flowers, but no more. But although I would be mad to suggest the resurrection of many old gardening techniques — not so long ago one eminent writer was campaigning on behalf of the hot bed as an aid to propagation — growing bellis for cutting seems a worthwhile notion, especially as you only need a few plants.

For bedding the 'Goliath' strain is now most commonly used, and the flowers are indeed large and only decline in size a little as the later flowers appear. Only 6in (15cm) high, the flowers are 3in (7.5cm) across and come in various reds and pinks plus white. For me the proportions are all wrong and they do look a little tatty as they go over. 'Neibelungen' looks better proportioned as the tightly double flowers are carried on 9in (23cm) stems. This is the one to go for for cutting.

But some of the smaller-flowered varieties with densely tight heads such as 'Kito' in cherry red are quite delightful. Look out too for the 'Pomponettes' reaching just 6in (15cm) at most in red, pink and white; the flowers have an attractive yellow eye when they first open. They tend to flower a little on the late side — May to July — so finding companions for them may be difficult. Try a dwarf daffodil such as *N. canaliculatus* or better still use them in the traditional way, as edging along paths. The summer plants can go in behind them and when they are removed later the alyssum or begonias will be poised to fill the space.

If you can bear the chore it pays to dead-head these daisies regularly. It is quite simple, a swift tug pulls the head off leaving the stem which is quickly hidden. This will ensure that the flowers keep coming, but more to the point for the gardener for whom any work spent on the lawn is simply time withdrawn from the proper business of gardening — looking after the flowers — you will have a lawn full of daisies if you do not. All that time sploshing on weedkiller will be wasted.

Two more points. You can, if you are sufficiently perverse, buy seed of the meadow daisy, described in the glossy catalogue as the 'single white daisy of lawns and meadows'. This is all very well if you want a meadow to encourage wildlife but not if you are a neat-and-organised gardener who values sanity. I once found a delightful form with bright yellow leaf veins growing in lawn at the botanic gardens at Glasnevin in Dublin. Risking nationalist wrath I extracted it with my penknife and potted it up at home. It was very pretty, had the happy habit of hardly ever throwing any flowers and was sadly burnt to a frazzle the following spring when the automatic greenhouse watering system packed up while I was away. Look out for another.

**Brachycome
(Swan River
Daisy)**

An Australian member of the daisy family, which grows naturally around Perth, brachycome is a delicate plant carrying vast quantities of flowers. Blue and white varieties are found in the wild but the colours available in mixtures now include various blues and bluish purples, lilacs, pinks and white. A very impressive rich purple, 'Purple Splendour', is also available as a single colour. The attractive characteristics of this pretty plant are the constant profusion of flowers, which cover the bushes, the fine delicate foliage, the scent which hangs on the still air delightfully and the fact that they are so easy to raise. They make good long-lasting cut flowers too.

Sow them at about 60°F (12°C) in March, and grow cool. It pays to pinch the tips out to encourage the natural tendency to a rounded bushy habit. They are fairly tough so can be planted out in early May if hardened off well. Sow them outside, too, in April and thin them to about 6–8in (15–20cm) depending on the fertility of the soil. They prefer sunshine and a well-drained soil and make a dramatic and long flowering edging along drives. If left entirely to their own devices they may give up flowering as seed is set so try attacking the plants with shears after about six weeks of flower, give them a good soak and maybe a liquid feed and they will soon be bursting forth again.

Once there were a number of selections available in separate colours — 'Blue Gem', 'Mauve Beauty', 'Purple King', 'Summer Beauty', 'Blue Star', 'Snow Star', 'Rosea' — all now vanished. One seed company at least has thoughts of making some selections but sadly the cost — which results from the time it all takes — may be prohibitive and so maybe this is a job for the keen amateur.

**Brassica
(Flowering
Cabbage)**

Not of course grown for its flowers, rather for its brightly coloured foliage. This is another of those 'love it or hate it' plants and I am afraid that I belong to the latter group. If you take to them, though, they make very colourful plants and, if grown in the way they are intended to be grown, bring a very bright contribution to the garden in the autumn when everything else is tailing off. Most only colour when the night temperature falls below 50°F (10°C) and as the plants quickly reach their full size, if sown too early they deteriorate before the leaves colour up. Sow in early July in a cold frame or greenhouse, prick out into 3in (7.5cm) pots and plant out in August for colour right through beyond the first frosts. Much development has been done in Japan where these varieties are very popular and there are now even a number of F1 hybrids — an extraordinary waste of energy this may seem to the gardener crying out for a hardy aubergine, or totally reliable 100 per cent double stocks. Instead of which we get pink cabbages, and you cannot even eat them.

The heads are flat and the foliage is often finely cut — as much as a cabbage can be. The central part of the head, and sometimes the ribs of the outer leaves, may be red, pink, cream or white or a combination of the colours. There are quite a number of varieties to be found and no one seems to have universal approval. The F1 types are more expensive but come to maturity rather earlier and are more uniform than other types.

If you are giving some spare winter cabbage plants to a neighbour pop a few of these in for a bit of a surprise!

# C

**Calandrinia (Rock Purslane)**

Calandrinia is another of the small but select band of annuals that actually look right when grown on rock gardens — although the extraordinary colour always offends the purists. In the wild in Chile and Peru, many calandrinias are perennials but in the uncertain British climate they are best grown as half-hardy annuals; it has even been suggested that they be grown as biennials but this seems an unnecessary inconvenience.

There is only the one species usually available and it is probably the showiest of all. *C. umbellata* grows 6in (15cm) at most and the leaves are narrow and greyish green in colour. The flowers are a startling magenta shade and so their neighbours need careful selection. Avoid purple and scarlet at all costs but opinion seems divided on orange — though not in this book. Sow seed in March at about 50°F (10°C) and prick out into trays in the normal way. Growth can be very slow at first and damping off is a special risk. Keeping the compost on the dry side is the best precaution for this plant and when it goes outside dry sunny conditions will give the best show. In the south-west of Britain self-sown seedlings may appear in autumn and flower early the next year but in most parts of the country the summers are not good enough for seed to be set.

**Calceolaria**

Not the commonest of bedding plants these days and rarely seen in private gardens. Most of the development work has been done to produce new varieties for growing as pot plants and the 'Anytime' series, which will flower in just over four months whenever you sow it as long as the temperature is right, is the culmination of this work. The massive promotion of marigolds has tended to push other yellow-flowered bedding plants from the public eye and this is very sad. Marigolds are all very well, but the unusual form and more delicate habit of the calceolaria has a lot to offer.

Once they were grown from cuttings and overwintered in warm greenhouses for planting out in spring and a very airy and soft display was created. Fortunately the situation has been somewhat saved by Floranova, an independent British plant breeding company who are not only independently managed but who also pursue an independent and innovative line in the plants they choose to work with. They have produced a splendid seed-raised calceolaria, 'Midas', which is pure yellow and which flowers for months on end. Its other great advantage is that it can be raised in a cool temperature. It can be sown in a frost-free greenhouse in early spring, planted out in May and it will flower right through to the frosts. If you have no greenhouse, try sowing in a cold frame in late April and planting out during June. Good with red verbenas, *Tradescantia pallida* 'Purpurea', cut-leaved perilla, or for a really fiery display, what about a coleus like 'Red Monarch'. 'Sunshine' is another good variety which flowers slightly later than 'Midas', has very good weather resistance but produces rather irregularly-shaped plants. 'Golden Bunch' is paler and more compact.

Calceolaria seed is extremely small, some of it even smaller than begonias, so you need to take great care when sowing it. The begonia routine will do very well; it is crucial to ensure that the surface of the compost does not dry out in the early stages otherwise your minute seedlings will vanish.

**Calendula (Pot Marigold)**

Also known as the English marigold it is not actually English or even British but originates in Southern Europe. It escapes from gardens and sometimes persists for a year or two in waste places but no more. In the wild it reaches 3ft (90cm) but in cultivation the varieties range from about 9in

(23cm) to 2ft (60cm). All those in cultivation are forms of *C. officinalis* and although there are other species around — one or two of the shrubby ones might make good bedding perennials — it is the inherent variation in the one species that has led to such a range of good varieties. Flower colour varies from chestnut to lemon, but without good reds plus a creamy apricot, which is the nearest to white. They are hardy annuals with interesting curved seeds, usually produced in sufficient numbers to keep the cost down as well as overpopulate your garden for a year or two after you first grow them.

Sow in March in the open and they should flower by May but if you really like them be prepared to sow again later. Even if dead-headed meticulously they tend to fizzle out. Mildew does not help and all varieties tend to go down with it sooner or later, even the new F1 hybrids. Spray with one of the systemic fungicides every couple of weeks and you should forestall attack but sooner or later the plants will succumb. The French variety 'Anagoor' is probably the most resistant but is rarely available to home gardeners. If you want to raise them indoors then April is quite early enough to get under way.

*Figure 7.2* Calendula *'Indian Prince'*

As to soil and position, they must be amongst the most amenable of all. Preferring sun and a fertile but not too wet soil — too much moisture and you will end up with a big green mound covered in little orange specks — they will make a show in most situations. Dwarf varieties are good edgers and good for containers and also make pretty winter and early spring pot plants. The medium and tall ones will pack a real punch of dramatic colour — so do not put them too near windows or where you sit, and they also make superb cut flowers and their colours are like no other. Many special varieties have been bred for this particular use, some excellent ones from Japan, and the appearance of dark mahogany shades on the backs of the petals in varieties like 'Indian Prince' together with especially clean and elegant forms make flowers no flower arranger — or those of the stuff-them-in-the-vase school — would want to be without.

Amongst the dwarf varieties 'Fiesta Gitana' was a Fleuroselect Bronze Medal winner in 1977 and grows to a maximum of 12in (30cm), often 9in (23cm). It branches strongly often producing a plateau of flowers initially which becomes more rounded later. The flowers are more or less double and the mixture contains six colours but the orange, yellow and various slightly insipid shades, even with the dark brown eyes, do not make a very impressive display. The orange and yellow are available separately and make more elegant alternatives to the ubiquitous marigolds. Strangely, only the orange won an award in the 1982 RHS trials. Unfortunately the very dense habit tends to encourage mildew. They are good in window boxes.

'Mandarin' was the first F1 hybrid. The plants reach about 10in (25cm) and the flowers are big, almost 3in (7.5cm), in bright orange. It grows rather taller in good conditions and has a sibling, a very pale, pinky yellow-flowered variety called 'Apricot Sherbet'. Both these two are likely to disappear in the next year or two. 'Anagoor' is rarely available but has 3in (7.5cm) orange flowers with dark centres on a plant just 9in (23cm) high.

In the slightly taller range, 'Radio', an old variety used for cut flowers in the 1950s, is still popular. An 'Extra Selected' strain has big brilliant orange quilled flowers in the form of a cactus dahlia. Superb for the border, long-lasting and prolific as a cut flower, it reaches about 18in (45cm). 'Neon' is a relatively new variety reaching about 15in (38cm), with flowers about 2½in (6.5cm) across and a semi-incurved form. The petals are yellow on the upper surface and deep orange on the back making a very dramatic combination.

For a tall variety try 'Art Shades', a standard variety growing over 2ft (60cm) and containing a mixture of colours including orange, apricot, yellow, cream and so on. 'Kablouna' has large, crested flowers in various yellow and orange shades many with dark centres. 'Kablouna Gold' is available separately and is gold with an orange crested centre and a dark eye. It was Highly Commended in the 1982 RHS trial. 'Orange King', darker than 'Radio', is a good double orange reaching about 2ft (60cm) which won an Award of Merit in the RHS trial.

Three varieties from Japan are also available. They are distinguished by their upright habit and repeated basal branching together with generous flower production and interesting shades. 'Early Nakayasu' has almost fully double orange flowers, 'Muraji' is deep orange and fully double while 'Yashima' is the most vigorous and is deep orange with a dark centre. 'Pacific Beauty' used to be a good tall type but stocks have deteriorated now and they are less reliable.

All the taller varieties benefit from support and in the border hazel pea sticks are still the best bet. Cut flowers grown in rows are best supported by horizontal netting.

## Campanula (Bell Flower)

There are two very distinct types that fall within the scope of this book although there is also a great range of others which are either herbaceous or rock plants. For summer baskets and other containers there are the varieties of *C. isophylla* and there are also the biennial Canterbury bells. For many years *C. isophylla* was grown by parks departments and a few keen gardeners as a summer-flowering greenhouse plant which was renowned for the fragility of its branches; a slight tap and they broke. They were trailing plants, sometimes making quite long, loose stems, with silvery foliage and pretty blue or white bell flowers. The plants were tender and had to be overwintered in a frost-free greenhouse.

Then in 1983 two new seed-raised varieties were introduced — 'Krystal White' and 'Krystal Blue'. These are sown in January and make good-sized plants for baskets and other containers by May. Unfortunately the seed is extremely fine, like begonias, and is also expensive so when you look at the transparent inner packet it appears as if you have paid a fortune for a few specks of nothing. But the plants, which will bloom from June onwards, are less brittle than the older types, flower profusely right through into the autumn and are a little more compact than the old types as well. More to our point, they thrive outside and in just a few years have become widely used, especially in baskets. Of course if you keep a greenhouse frost-free all winter they will keep on flowering as long as the light is reasonable and you will still be able to take cuttings in spring for fresh plants; but if you can raise begonias from seed, these are no more difficult. A new compact variety, 'Stella', has recently appeared.

The blue and white forms make a lovely basket planted together with nothing else and they are also good with any other fairly compact plants like fibrous begonias — the reds and pinks associating well with the blue and white.

*Figure 7.3*
*Canterbury bells*

Canterbury bells are altogether different. They are genuine cottage garden plants to raise by sowing outside in late spring and transplanting to their flowering homes in the autumn. There are two types — the 'cup and saucer' type has an enlarged calyx coloured the same as the flower while the standard variety has a small green calyx. The plants reach 1¼–4ft (0.4–1.2m) and the flowers are large and of a genuine bell shape with a slightly recurved lip. The plants are often of interesting pyramidal shape with flowers all round, on all sides and along most of their height. 'Dwarf Bedding' reaches 15–18in (38–45cm) and comes in blues, mauves, pinks and white while the 'cup and saucer' types and the ordinary mixed come in a similar range of shades but can reach 4ft (1.2m). The dwarf types make excellent cold greenhouse pot plants and are best flowered in 5in (7.5cm) pots.

**Celosia (Feathered Cockscomb)**

Celosias are usually grown as pot plants and the crested cockscomb, one of the ugliest of all 'flowering' plants, is especially popular. Fortunately, this will not succeed outside but its more elegant relation, the Prince of Wales' feather it is also called, with the tall, feathery plumes is a different matter. This too is often grown as a pot plant but in the warmer areas it is quite happy raised by the normal half-hardy annual technique. Although only a small plant it is nevertheless a hungry feeder and it is wise to start liquid feeding a little earlier for this plant than most half-hardy annuals. It also pays to prick out into 3in (7.5cm) pots to make sure the maximum amount of plant food is available.

There is not a great range of varieties. 'Apricot Brandy' is a good, basal branching sort whose colour is made clear by its name. It needs careful choice of companions but *Cineraria* 'Silverdust' is good and it looks well in front of a dark-leaved dahlia like 'David Howard'. 'Fairy Fountains' at around 10in (25cm) is a good mixture of reds, oranges, salmon, yellow and pink and even shorter is 'Red Kewpie' which is bright scarlet and makes an interesting alternative to salvias in warmer areas. The recent All America Selection medal winner, 'Century Mixed', has a good colour range too, especially in the yellows.

In cold, damp and windy parts of the country, it is probably wiser not to risk them, although as pot plants in the greenhouse they will still be excellent value.

**Centaurea (Cornflower)**

The cornflower was once one of the commonest, and the most colourful, of cornfield weeds with Parkinson writing in 1640 that it was 'furnishing or rather pestering the cornfields'. But its life cycle was against it as the majority of the seeds germinated in the autumn, overwintered as rosettes in the field and then flowered in summer. As soon as the fields were enclosed and drilling became more common the standard of weeding rose appreciably and the decline of the cornflower set in. At first sight the tendency to overwinter cereal crops would appear to make the re-appearance of the cornflower a possibility but the standard of chemical weed control is such that, even if viable seed were to be turned up by deep ploughing, and the seed does last a long time in the soil, it would be wiped out by the regular weedkiller applications. But as you might expect this is one of the best of annuals for late summer sowing to flower in spring when the plants are far more prolific than those sown in spring.

The flowers, also known as bluebottle and bachelor's buttons, are bright blue with a purple centre in the wild form but now a variety of shades is available. There is 'Blue Diadem' with big deep blue flowers up to 2½in (6.5cm) across and 'Red Ball' in dark pink shades. The mixtures such as 'Polka Dot' contain blues, various dark reds, pinks and some pale almost white shades. In 'Frosty' there are varying degrees of white or pale flecking. 'Double Mixed' is often seen but tends to contain only blues and whites with the occasional dark purple. But at 2ft (60cm) it is a good height for cutting, although this variety can finish early if not cut. 'Jubilee Gem' is half the size, has deep blue flowers and very dense growth. They are all superb cut flowers and are best grown in a row in a space in the vegetable plot where they will not get in the way. They prefer sandier soils and plenty of sun and should be sown in August or September in rows 18in (45cm) apart and thinned out to around 15in (38cm). They may need staking in spring when the flower stems begin to run up. The first flowers should appear during May and they are best cut when they are fully open. By sowing in spring as well you should get flowers until September.

The sweet sultan, *C. moschata*, is not often grown these days but is nevertheless an attractive cut flower. It has large, fine, fluffy heads usually in white, yellow or shades of purple, on plants up to 2ft (60cm) high with deeply toothed leaves. As you will doubtless guess from the name the flowers have a sweet scent. Standard hardy annual treatment suits them, thinning to about 9in (23cm) between plants with rows 12in (30cm) apart.

*Centaurea gymnocarpa* is grown for quite different reasons. This is a large, silver-foliaged plant reaching 18in (45cm) if treated as a half-hardy annual and a very useful step up from *Cineraria* 'Silverdust' and the like. It needs to be sown early to make a good-sized plant but it will survive an average winter outside in many parts of Britain and so can be increased from cuttings in spring. The foliage has a slightly metallic sheen and is finely cut; in late summer you may also get small, double, dark pink flowers, which can be left or cut out low down according to your fancy. Lovely behind 'Chiffon Magic' petunias and a good plant for large containers where other plants can scramble through it.

**Centranthus (Soldier's Pride or Red Valerian)**

Native in Central and Southern Europe and also in North Africa and Asia Minor, the red valerian was a cottage garden favourite for many years. Its clusters of small red flowers on bright, clear green shoots and its tolerance of drought and neglect still make it popular. By the end of the nineteenth century it had left gardens and in Kent had become established on chalk banks and railway embankments. Now, especially in the south west of Britain, it is frequently seen on old walls, dry banks, cliffs and in any waste places where the soil is on the dry side. The most common colour is now rather paler than the strong scarlet that led to its military name, this paler version being more vigorous.

You will appreciate from its tendencies in the wild that it likes dry, well-drained soil and preferably one which is on the limy side. You might also guess, from its vigorous naturalisation, that in the garden it is a pest. My father, having brought home a few seeds from a brilliant plant growing on a West Country wall, spent two years enjoying its brilliance and sweet scent and the next ten years trying to get rid of it. Seedlings sprang up everywhere and the habit of the stems breaking at ground level and the roots re-shooting was singularly exasperating. Now the garden is without it and Dad's enthusiasm for gardening back at its usual low ebb!

There is a mixture available now which includes various reds and pinks plus white and if you want to take the risk the seed can be sown where it is to flower and it will soon be doing its stuff. If you want to grow it in walls where it will stay rather smaller than its usual 2–3ft (60–90cm), then get up on a ladder with a few seeds and small watering can. Find a few crevices, give the area a good soak and then tuck the seeds in to the cracks. A more haphazard method, which also seems to work, is to simply throw a handful of seeds gently at the wall and hope for the best. But be ready to hoe them off from the border underneath.

**Cerastium (Snow in Summer)**

Until recent years no one thought of using snow in summer as a bedding plant. The truth is that few people think of it as a bedder now but as a plant for containers it really is useful. It is the silvery foliage of course that is the main attraction especially as it does not usually provide much of a show with its white flowers. It needs no more than the usual half-hardy annual treatment and can go into baskets or window boxes as about the only compact silver trailer. Try it with *Campanula* 'Krystal Blue' or 'Krystal White' or indeed the mixture of the two and it is also good with the yellow-leaved *Pyrethrum* 'Golden Moss' — 'Golden Feather' is a little too tall; the silver and lemony yellow are excellent. 'Yo-Yo' has slightly smaller leaves than the type and is relatively new on the scene.

**Cheiranthus (Wallflower)**

The first thing to sort out about this group, probably the most important of all bedding plants to the gardener, is what is a cheiranthus and what is an erysimum. Of course, you may wonder if it matters and this is open to question but given that seed catalogues have occasional aberrations and quirky habits in this area it deserves elucidation.

Botanically it is quite simple. Both have nectaries at the base of the stamens. In erysimums they more or less encircle the bases of the outer stamens and also appear on the outside of the inner ones. In cheiranthus they appear around the base of the outer stamens only. Mind you, this botany has little or no use when it comes to growing plants. Most wallflowers come under cheiranthus — 'Fire King', 'Primrose Monarch' and the like, all appear under this head and some of the perennial ones like 'Bowles Mauve' (surely one of the most underrated of *all* plants) sometimes appear there too. Under erysimum come the Siberian wallflowers — though their connection with

Siberia is far from clear; one authoritative work sums up this little geographical problem by saying that the plant's 'status and provenance are obscure'. Yes. Most of the other perennial wallflowers also come under erysimum. To be practical the division is most conventionally between 'wallflowers' and 'perennial wallflowers'. However, the conundrums do not end there. All wallflowers are perennial and the fact that on dry walls they make old and spectacular plants is testimony to this. But some are grown as perennials and others are grown as biennials, and that is the important division.

But, and I am sure you have been seething on behalf of the geranium and the petunia since the start, what makes the wallflower so special? First, there are few enough good plants to flower in the spring — only pansies and polyanthus have the variety of colour. Secondly they are so easy to grow, you do not need a heated greenhouse or even a frame, just a space in the garden. Thirdly, they are tough and put up with all but the most ferocious of weather. Next, they are cheap and, last, the scent — which can be one of the purest horticultural delights. The fact that few people grow their own is very sad as the limp specimens squelching in the buckets of slimy water in draughty markets rarely come in any colours other than 'Mixed' and 'Blood Red'. Not only can you grow plants in a variety of sizes and colours if you grow your own, but they will almost always survive when planted out in the garden.

Anyway, having got all that off my chest what about growing them. This too is a matter of controversy. Some say, do not sow until early July, others will say June is too late. I started by always sowing in May and I have got steadily later. Here is the crux of it. Try and plant out young plants that are as large as will survive in your particular garden. And that means sowing as early as will produce these plants. In mild, sheltered gardens where the soil is good you can happily sow in May, and have big plants to put out in the autumn which will produce a grand display. If you live in a cold, and particularly a cold and windy spot — and I have gardened in the middle of a Fenland sugar beet field so I know all about wind — then small plants are essential. This leads to an extra advantage of growing your own. As the poor plants will not have to be pulled from the ground by the leaves and stood in a bucket for a week before being taken home by the eager gardener, you can afford to use plants which are slightly larger. The soil which I hope you will keep on the roots will give them a far quicker start in life and they will bear their winter torment with greater resilience.

So, growing your own. The usual biennial treatment will serve ideally with the vital proviso that the soil be free of clubroot. Unfortunately wallflowers, like a number of unexpected members of the cabbage family, including stocks, suffer from this dreaded disease. Gardeners who, quite sensibly, grow their young wallflowers on the vegetable plot do not always remember to include them with the cauliflowers and the sprouts in the brassica rotation so there is always the danger of clubroot.

The dwarf types can look good in mixtures, especially if planted in a bed on their own or in large spaces in the border. But you will need to put in quite a few plants so that any local unevenness of distribution is balanced over a broad area. The 'Bedder' series is the best bet and there are far more colours in the mixture than just the four available as single colours. Gold, scarlet, orange and primrose are available separately and for small gardens make lovely combinations with daisies, forget-me-nots and single-coloured polyanthus. Try 'Primrose Bedder' behind 'Blue Ball' forget-me-nots or the lemon selection of the 'Pacific Giant' polyanthus.

There is a much broader range of colours available in the taller varieties which reach about 15–18in (38–45cm) and are especially good in the darker shades with 'Blood Red' a deep velvety red, 'Vulcan' a deep crimson and 'Fire King' which is a brilliant, slightly orangey scarlet. These can all be used behind their paler counterparts in the 'Bedder' series and of course with pale tulips. Orange with 'Blood Red', white or yellow with 'Fire King'. The other colours especially worthy of growing are 'Primrose Monarch', 'Cloth of Gold', 'Ivory White', a rather creamy shade, and 'Purple Queen'. The dark purple almost black tulip 'Queen of the Night' is especially good with the paler colours. If you have a large space, try growing four or five different colours of the taller types towards the back of the beds with the 'Bedder' series towards the front.

An interesting way to experiment with spring-bedding combinations is to plant out a bed of mixed wallflowers with a selection of mixed tulips in an out of the way corner — you get a horribly spangled effect but it will show you which particular colours of tulips go with which wallflowers.

**Chrysanthemum**

Probably the most brilliant and easy of hardy annuals, all gardeners should be forced to grow them! They are naturally long-flowering and with a September sowing and a late spring sowing too they can be in flower from spring to autumn. They make excellent cut flowers producing lots of long stems with a glittering array of colours and they last very well in water too. In borders they are very easy and if only they were more commonly available in single colours my garden would be full of them. The varieties that we grow outside now are mostly hardy annuals although some of the perennials can be persuaded to flower in their first season if sown early. The named perennials can be grown from cuttings and used in bedding schemes too, although their late flowering can disrupt the planning of other displays.

The annuals are derived from a number of different species and it makes sense to go through them one by one, starting with the, more or less, British one. The corn marigold (*C. segetum*) used to be a very common weed of cornfields and its habits in that situation set the tone for the growing of the whole group. It is probably native to the Mediterranean but arrived in Britain at an early stage and has now spread all over Europe as far north as Norway and has even turned up in the Americas. It has a distinct preference for acid soils, and what is more, acid soils which are not too heavy — you do not find it growing naturally on London clay. Modern agricultural practice has of course reduced it dramatically. This is partly the result of weedkillers, but the foliage is rather waxy and weed sprays do not have as much effect as they might. Liming has had a far greater influence. Nevertheless, I have seen it in sheets of gold on Benbecula in the Outer Hebrides.

The height of the plant varies from 12–20in (30–53cm) and the daisy flowers are about 2in (5cm) across and entirely gold. Not a blockbuster of a plant but very pretty and ideal in the softer, rather tumbling cottage style of gardening. There have been few named varieties available recently but one called 'Prado' has appeared. The flowers are up to 3in (7.5cm) across and the disc of the daisy is reddish brown while the petals are gold. Very impressive.

*C. carinatum* is originally from Morocco and even in the wild the characteristic ring in a contrasting colour around the disc is present. Since its arrival in Britain at the end of the eighteenth century many cultivated forms, including doubles, have arisen and there are a number of mixtures now available. They get the standard hardy annual treatment and will succeed well if sown in the autumn as long as the soil is well-drained. Try a sowing in May or early June to give vigorous plants in the autumn. They may need support from pea sticks in rich soil. 'Court Jesters', Highly Commended in the 1985 RHS trial, is clearly the best variety for the sheer brilliance of the colours — white, scarlet, maroon, pink, orange, yellow with broad orange rings. You have never seen such a cheerful show. If you insist on growing another variety try one of the other single mixtures with names like 'Special Mixed' and 'Merry Mixture', but steer well clear of the horrid doubles which have very little appeal except that having grown them you will be convinced of the futility of the exercise and will then doubtless put more faith in the veracity of this text! Oh, one more, 'Cecilia'. The flowers are over 3in (7.5cm) across, on plants 3ft (90cm) high, and are white with a yellow centre. It won an Award of Merit in the 1985 trial.

*C. multicaule* is a wonderful, low-growing plant sometimes described as a half-hardy annual but quite happy sown outside if left fairly late. It reaches only around 9in (23cm) and spreads out very well sideways. The foliage is rather fleshy and the bright yellow flowers are about an inch (2.5cm) across. The variety 'Gold Plate' is about the only one you will find these days. The flowers of the variety are rather larger at 1½in (4cm) and some are double and semi-double. Great in window boxes, the corners of borders and worth a try in hanging baskets too.

*C. padulosum* is a sort of miniature marguerite that grows to about 9in (23cm) and is covered with 1½in (4cm) white daisies on rather stiff foliage. Very dainty and useful in gaps but do not let that rather dismissive remark put you off. Supposedly a perennial but I have always wanted it out to make way for something else.

In both good and bad summers, one of the best of all annual chrysanthemums is a variety of *C. coronarium*, a tall plant originally from the Mediterranean. 'Golden Gem', also Highly Commended, is the one you usually find and this grows to about 18in (45cm) with bold yellow daisies almost 2in (5cm) across on clear green foliage. Each new flower opens just a little above the last fading one so the plant always looks good from June well into September and longer. The growth is very dense and weed proof and the whole plant eventually takes on a very rounded appearance.

Producing F1 hybrid chrysanthemums is, for genetic reasons which I need not go into, rather tricky — far more difficult than dealing with geraniums as their awful proliferation makes only too clear. However F1 hybrids have been created and they are said to do away with the necessity of overwintering roots from one year to the next. Have none of it. You will have to sow them in January or earlier, when heat is at its most expensive, they will not flower until mid-summer and the flowers will be 'quite pretty'. The seed will also cost you five times as much as the 'Court Jesters'. Buy a few plants of Pennine sprays instead. They will usually survive outside over the winter and if they die buy a few more the next year. It will probably not cost you much more in the end and the results will be stunning.

**Cineraria**

Under this heading we find — confusion. First of all we are not talking about the cinerarias that are so often grown as pot plants in the first half of the year. There are places, such as Hampton Court, where plants in flower are slotted into the border to fill a gap in the cycle of colour but even I am not going to suggest such a thing, especially as home gardeners do not always find them easy to grow. It is the foliage varieties that concern us under this heading but I must say that they occur under this heading not through a botanical aberration but because three-quarters of the catalogues list them under this name even though they are strictly varieties of *Senecio cineraria* also known as *S. maritima*. I say varieties, plural, although one, 'Silverdust', takes almost 100 per cent of the sales.

'Silverdust' is a more or less hardy perennial sub-shrub which is raised as a half-hardy annual for bedding. It has rather solid, broadly lacy leaves with dense almost white dusting giving a very bright effect. It only reaches about 9in (23cm) in its first year and will not flower. In most parts of Britain it overwinters happily, especially in well-drained soil and can reach 1½–2ft (45–60cm) in its second year before rather disgracing itself and throwing out long, straggly flower stems carrying loose heads of little yellow dandelions. You can cut them off, of course, and cutting back in May will help improve the form. It should be said that if you leave this plant to overwinter you can raise it from cuttings the following spring but that they will not necessarily fight shy of flowering.

But there are quite a few good perennial silver plants so this one really ought to be kept as an annual. The other variety that is sometimes listed is a more recent one called 'Cirrus' which has oval, slightly wavy, notched leaves and which grows just a little taller in its first year. The leaves start off a silvery green shade and get slowly less green and more white as the season goes on. Both are excellent although 'Cirrus' is rather expensive. One more is occasionally found which is taller at about 15in (38cm) and that is 'Diamond'. This has more or less oak shaped leaves. It is not quite as white as 'Silverdust' and makes a good substantial plant for a little further back in the border but it tends to be rather variable in form and colour.

In public parks you will see this plant, especially 'Silverdust', used with scarlet geraniums like 'Grenadier' and strong blues like Lobelia 'Mrs Clibran'. It is preferable to alyssum in that situation as it actually lasts right through the summer without burning out. This is very dramatic if you like that sort of thing and you could also try it with 'Inca Orange' marigolds and the lobelia. For something a little less in the blunderbuss style of bedding use it as an edging to a very pale lemon such as 'Solar Sulphur' marigolds or 'Chiffon Magic' petunias with clusters of the rather sharp, dusky blue *Salvia farinacea* 'Victoria'. The slightly taller 'Cirrus' can be mixed with a really dark dianthus like 'Crimson Charm' or again with one of the slightly larger ageratums like 'Blue Champion'.

And lastly, if you see a plant on the coasts of southern England that reminds you of this one, it is the very same. Although it came originally from the Mediterranean, it has escaped from the gardens and now finds itself growing on cliffs and shingle.

**Cladanthus**

Ever since I weeded round this distinctive little annual at Kew many years ago this has been one of my favourites. It is a member of the daisy family and has bright yellow flowers up to 2in (5cm) across. Its manner of growth is one of its most endearing characteristics. After the first flower, a number of side branches are produced in a ring around the flower as it fades. Each of these carries a flower in its turn and the effect is of an orbit of yellow satellites around the dying flower. The growth is rather stiff and as all these rings of stems intercross, an unusual and rather elegant effect is created. The foliage is fine and ferny and the

*Figure 7.4*
Cladanthus
arabicus

flowers also have a delightfully pleasant scent; by the time the plant reaches 2–3ft (60–90cm) it is a pretty and charmingly perfumed plant. The fact that it flowers for so long is an extra bonus. Leave a few old heads and the chances are that you will get some self-sown seedlings the following spring. *C. arabicus* is the only one you will find.

**Clarkia**

Clarkias are perfectly hardy and good for spring or autumn sowing. They support themselves happily even though they are likely to reach 2ft (60cm), they flower for weeks and weeks and the colours, even though they are bright, spare us the garish brashness of salvias. Thinned to 9–12in (23–30cm) they make stiff, branching plants, ideal for cutting. They fit in well amongst shrubs and permanent plantings and are not so blatant as to overemphasise the face that the shrubs are looking a little dowdy. Almost any soil will suit them though being Californian they like some sunshine. If you hunt about you can find them in single colours: 'Albatross' — white, as you might expect, 'Brilliant' — double carmine, 'La France' — double salmon, 'Orange King' — double reddish orange and 'Purple King' — double deep purple. All are worth growing. For a mixture go for 'Monarch', raised by Hursts, which includes four shades of pink, purple and white. It is tall, especially when grown on good soils and can reach 4ft (1.2m) but the flower stems will take up half that height. Try it fronted with 'Fairy Mixed' candytuft.

These are all varieties of *C. elegans* but a couple of others are grown too. *C. pulchella* is rather shorter, about 12in (30cm), and more delicate and is usually found in mixtures of reddish, pinkish lilac and white shades — the named varieties that were once grown have all vanished and they are not now seen as pot plants either. *C. concinna*, another from California, is even shorter at about 9in (23cm) and has rather frilly flowers. 'Pink Ribbons' with its deep flowers is about the only one you will come across and is an excellent window box plant.

**Cleome (Spider Flower)**

Strangely constructed flowers each with three or four petals at the top, a long style and five or six stamens creating a spidery appearance. The plants reach 3½ft (1.05cm) and the flowers are clustered in tall terminal heads. The divided leaves, though not the stems, have spines running along the undersides of their stems and veins. Standard half-hardy treatment suits them and they are happy in a well-drained though not impoverished soil in full sun. The pink is the commonest seen and there is a good mixture, 'Colour Fountain'. Single colours — white, violet, carmine — are also available as well as the pink. All are good cut flowers and also associate well with any other plants of softer shade as well as taller silver foliage plants.

**Cobaea (Cup and Saucer Plant)**

A vigorous and even rampageous climber from the southern United States and Mexico, you can guess the sort of conditions in which cobaea thrives — it does like it warm. In Britain it can be grown outside in the south and in sunny and sheltered spots further north but in colder districts will do best in a conservatory. Try it in a greenhouse or conservatory in the south and by August you

114

will hardly be able to get in the door. And what is more, the big bell-shaped flowers are on such long stems, up to 12in (30cm) in warm conditions, that growing towards the light they are quite hidden from view by the mass of foliage. Outside they grow from 6–10ft (1.8–3.0m) depending on the conditions and need some trellis or wire for support. The flowers are interesting in that as the buds open they are rather creamy in colour, then they quickly turn green and finally end up a strong purple which is really their most long-lasting shade. In the wild they are pollinated by bats but in spite of sitting up late with a glass of scotch and a keen eye I have never seen it happen.

Being fond of warm weather and sheltered spots they need a long growing season so in less clement climes they are best sown fairly early — say early March — and they need a high temperature. I find that it pays to soak the large flat seed overnight in warm water, in a saucer near the central heating boiler for example, and then sow the by now slightly oily seeds on edge, singly in 2½in (6.5cm) peat pots of peat-based potting compost. Once they have germinated the temperature can be reduced to about 60°F (15°C) and when they have made a few inches of growth a split cane can go in to support them. Pinch the tops at about 8in (20cm) and when the roots start to appear through the sides of the peat, pot them on into 5in (12.5cm) pots of the same compost. By this time the temperature that you maintain for everything else in the greenhouse will suit them well but they must be hardened off thoroughly before planting out. A good soil helps them to get going quickly and they need it later as they often produce a great deal of growth. Old growing bag compost is, as usual, ideal and a hole about 1ft (30cm) square and as deep is fine. Fork the compost into the bottom and mix it with the garden soil. A good soak is vital and protection from the wind helps too.

This plant is so quick and so dense that it makes a good screen and is well suited to growing alongside the patio to provide shade in the heat of the day. Annual climbers are not exactly common but try canary creeper sown in the soil in April to mix with it later on as they scramble together over the trellis. In nature this plant is a perennial and will overwinter in a frost-free greenhouse. If cut back hard in late autumn it will burst forth with astounding vigour the following spring. Generally though, I think this is as unrealistic as growing runner beans as perennials.

There is a variegated version which is slightly less vigorous but is rarely seen in catalogues. It does not come true from seed so must be overwintered and raised from cuttings of side shoots taken in spring and rooted in a sandy compost. The lovely white-flowered form has reappeared in catalogues so give it a try.

**Collinisia**

*C. bicolor* is a much neglected annual in the antirrhinum family with an ideal combination of characteristics. It is hardy and the flowers are very pretty with the upper lobe being white and the lower a rosy purple. It only grows to about 18in (45cm) and is happy in shade especially as the soil needs to be a little on the moist side, though not claggy. It seeds itself about the place happily but never in such an aggressive manner as to arouse the gardener's temper and will flower early in the spring from seeds dropped the previous summer. Years ago, this, like so many other annuals, was available in a range of colours, especially pure white and strong purple, but these have disappeared. It was also grown as a pot plant though I suspect this treatment was given more often to some of the other species which are now only very rarely available.

Grow collinsia in mixed borders where, once you have it, it can be left to go about its business. Learn to recognise its seedlings and fork out all those that are liable to be troublesome, leaving just a few to thread their way between shrubs and peep through more or less at random — they are not heavy enough to do any harm.

**Coleus**

Coleus have not prospered as bedding plants in recent years. Many new varieties have been introduced but they have usually been intended as pot plants rather than as garden plants. The result is that when planted outside these varieties grow very slowly and do not make the great balloons of colourful foliage that some of the old vegetatively propagated varieties do. This is a great shame for the colours available are often unique. The old types also produced extraordinary shades, matt orange ('Sunset'), soft red, edged grey ('Picturatus') and dusky purplish black ('Rois de Nois'). Sadly these varieties are almost never seen listed now, there are only the seed-raised types. The best advice, therefore, unless you come across these old types, is to buy seed of one of the more vigorous mixtures such as 'Rainbow Prize Strain'. Make sure to grow all the colours and then pick out the best shades and the most vigorous to keep on the windowsill as rooted cuttings for next year. Some of the smaller types like 'Wizard', 'Sabre' and, even more so, the 'Mini-Coral' series will survive most of the summer but at the end of September are still likely to be too small to make much impact. In 1985, 'Golden Wizard' made a plant only 4in (10cm) high by 6in (15cm) across by the end of September.

One interesting development has been the introduction of a variety intended specifically for hanging baskets. 'Scarlet Poncho' has slightly toothed, rather than heavily notched or waved, leaves in the traditional coleus nettle-leaf shape and each leaf is bright red with a narrow yellow edging. The plants branch well from the base and are not stiff and upright but lax in habit. It is rather a shocker, needs a warm spot and the basket must not be allowed to dry out. Very striking.

The usual half-hardy treatment suits them but they like a high germination temperature of 72–75°F (22–24°C) and if you can make sure that the surface of the compost never dries out, sow uncovered.

**Convolvulus**

Some people are rather alarmed at the suggestion that we grow bindweed in the garden, most have quite enough and a bind it really is. Some lunatics even suggest growing bindweed in hanging baskets! If you must grow such a thing then pick the sea bindweed, *Calystegia soldanella*, instead, it is far prettier. Better still do not grow it at all — unlike the greater bindweed it is self-fertile so even if you only have the one plant they will soon be jumping up amongst treasured alpines and in similarly irritating places. Instead, grow the hardy annual *C. tricolor*.

This is a very showy little plant even in the wild in Sicily, Spain and Portugal where it grows naturally. It grows about 12in (30cm) high, upright at first but then with a tendency to collapse. The flowers are trumpet-shaped and a very strong royal blue shade, with a yellow throat and a starry white band in between. The stems have an attractive red cast. They are happy in most soils that are not too dry and will put up a good show in partial shade. This is one of the best plants for attracting hoverflies, the larvae of which are the most voracious aphid gobblers we have so they are always worth a home just on that score.

There are a number of rather insipid mixtures and the uninspired appellation 'Mixed' is about as exciting as the colours. Thin pinks, watery blues, they all pale into insignificance alongside the 'Royal Ensign' which also seems to be known as 'Dark Blue' and 'Blue Flash'. A vibrant plant for the front of the border which in spite of being so vivid is never overpowering. The rich blue flowers are around 2in (5cm) across with a white feathery star in the centre and an orange yellow eye.

Good in tubs and window boxes it is best treated as a 'pop out here and there' plant rather than one which will make a more solid burst of colour. Try it with *Helichrysum* 'Limelight' for a love-it-or-hate-it combination or more soberly in front of white nicotiana.

The climbing types that used to come under convolvulus now appear under ipomoea.

## Cosmos

Cosmos are brilliantly coloured, half-hardy annuals which rival the annual chrysanthemums for the impact of their display. The foliage, which is finely divided into narrow filaments, is especially pretty and the vigour with which the flowers are produced makes them essential garden annuals. They can be sown inside in April, but are speedy growers so do not sow any earlier, or they can go into open ground where they are to flower at the end of April or early May. They are happy in most soils. A sunny, slightly sheltered site in a dryish but not impoverished soil is ideal although like so many annuals they will put up with most reasonable conditions. The flowers tend to be carried on long stems, so, if overfed, the whole display loses its impact and the plants also need support. They are quite likely to self-sow if the dead flower heads are left on and, if you practice the less organised style of mixed border gardening, dead-head the majority of the flowers to ensure a continuous display but leave a few to shed their seeds. They make splendid if rather unruly cut flowers and named varieties in separate colours were once grown for this purpose; sadly these are now no longer to be found. The different varieties vary in their heights, colours and uses so it makes sense to look at them individually. The varieties are based on two species from Central America, *C. bipinnatus*, which comes mainly in pinkish shades and white, and *C. sulphureus*, which of course comes in yellow plus orange and scarlet shades.

The best known in the first group is 'Sensation'. This is the most widely available mixture, it reaches about 3ft (90cm) with big single flowers up to 3in (7.5cm) across. Some see the colour range as rather restricted, others as finely toned but either way you will only find carmine, pink and white shades — although just about every nuance will be present. Ideal for cutting, it also makes bold yet soft groups in borders and self-seeds well. A fully double type has recently been re-discovered in a competition organised by a seed company although this is not an entirely pure strain. Selection is in progress and we may see a fully double type again before long. Needless to say, doubles were once quite common. 'Pyche' is a semi-double strain that only comes about 80 per cent true. The flowers are carried a little more upright than other forms. 'Sea Shells' is an extraordinary recently introduced variety, which comes in the same colour range as the 'Sensation' mixture and reaches the same height but has petals rolled to form hollow tubes instead of flat rays; a few may be two-toned. It looks quite bizarre and is best suited to more adventurous flower arrangements rather than garden display.

Some single colours have recently become available again and these include 'Purity', an invaluable plant; it is tall, 3ft (90cm), with pure white flowers — an unusual combination. 'Candy Stripe' has hideously coloured flowers, white with a dark purple rather variable picotee edge to each petal; alternatively the petals may be flecked or spattered with colour. Quite tasteless and if you see some in a garden I am tempted to recommend that you clamber into the border and heave them out.

In the generally rather smaller *C. sulphureus* group comes 'Bright Lights' which is slightly shorter than 'Sensation' at about 2½ft (75cm) and the flowers are rather smaller too. But they come in brilliant red, orange, yellow and other flame shades and are semi-double. Another good cutter, it makes a dramatic punctuation in borders. Still too tall for containers. 'Sunny Gold' makes about 18in (45cm) and is a semi-double, golden-flowered type; it can be said to represent the charm which totally eludes marigolds. It is very free flowering, with blooms up to 2in (5cm) across and for those gardeners who cannot bear the smug formality of so many marigolds this is worth a try. 'Diablo' is flame red, reaches about 2½ft (75cm) and is very useful in hot, fiery summer borders with red foliage dahlias, *Ricinus* 'Impala', perilla, scarlet dianthus and red-leaved, orange- or scarlet-flowered cannas. 'Lemon Twist' is again only 2½ft (75cm) a cool, lemon yellow shade. An 'All America Selection' bronze medalist has recently been introduced, 'Sunny Red', which as you might guess is a red counterpart to 'Sunny Gold'. The colour is quite as brilliant and at 12in (30cm) it is a few inches smaller than 'Sunny Gold', and a valuable addition to the summer garden flora.

**Crepis**

Some plants are neglected for no good reason at all and the delightful *Crepis rubra* is an example of a plant which just seems to have been overlooked. From the day I came across it in one of the herbaceous beds at Kew, I have always tried to have it in the garden. On the opposite corner of the bed was a white form but that is still not to be found, although the occasional white plant may turn up amongst the pink. The plant grows 12–15in (30–38cm) high and is very upright in habit. It makes a rosette of slightly puckered leaves, not unlike many of the coarser members of the daisy family, then the flower stems are thrown up and on the top of each, soft pink dandelions. The colour — do not take any notice of the *rubra* — is perfect and this is a classic plant for those pink, silver and sky blue schemes which have in many gardens taken the place of the red, white and blue. You will find one of the ageratums, say 'Blue Danube' or 'Ocean', the right shade. It is important to keep pulling the heads off as they pass over but leave a few on and harvest the seed for the following year. And if you ever come across the white version, beg a little seed from the gardener. But grow it at the opposite end of the garden from the pink or the seed will not stay true.

**Cuphea**

I was first struck by a cuphea — no it didn't jump up and clout me but nevertheless made a great impression — in the double red borders at Hidcote Manor in Gloucestershire. This was *C. miniata* 'Firefly'. It is a tender shrub, usually treated as a half-hardy annual, with long narrow tubular flowers towards the tips of the red stems. As you might expect, the flowers are a rather bright red, but not quite scarlet. It grows to about 12in (30cm) making a bushy plant which can be lifted in the autumn and will flower for some weeks in a greenhouse with frost protection. At Hidcote it was used with a number of other red-flowered and red foliage plants but I have seen it in a container with a trailing lobelia such as 'Sapphire'. Whatever you do, steer clear of the so-called red trailers like 'Red Cascade'. Their magenta-purple shades will clash horribly. The usual half-hardy annual treatment will suit it although if you overwinter a plant for its flowers a few cuttings can be taken in spring and will root without too much trouble. Also available is *C. ignea* at about the same height but the flowers are red with purple and white tips.

**Cynoglossum (Hound's Tongue)**

There is a British species, *C. officinale*, which grows in grassy places on dry soils, especially near the sea, and which was once used as a cure for stuttering. It is a grey-leaved biennial with dull reddish purple flowers usually appearing in mid-summer. A pretty, but rather leafy plant, it reaches about 3ft (90cm). In gardens the biennial *C. amabile* from Asia is most commonly grown, but grown as an annual. T.C. Mansfield in his book on annuals published shortly after the war is scathing about the variety 'Firmament', the only one usually available, which is compared to its wild progenitor and said to be 'more compact-growing and has flowers of a deeper blue, and is not better for either'. We, though, are stuck with 'Firmament' so we must make the best of it. It reaches about 15in (38cm), has grey green, rather downy leaves and carries vast myriads of intense blue flowers. There are not too many blue flowers amongst the summer annuals so it is perhaps surprising that this plant is not used more often. It does not have the blocky look of so many modern bedding varieties so is an ideal plant for the slightly unruly borders which are becoming increasingly fashionable. Unfortunately in especially dry summers its slight suscepti-bility to mildew is a problem but regular spraying with benlate every two weeks will solve the problem. It needs sowing early, say March, to make sure of a long flowering season but should be grown fairly cool. It is hardy so can go out rather earlier than most summer plants raised in heat, but this may not always fit in with the usual planting regime.

The alternative, of course, is to sow outside in the autumn, say September, although in wet soil and bad winters plants can rot off. If it is a plant that especially takes your fancy try sowing in autumn, pricking out into a well-drained compost and overwintering in a cold frame.

A true annual, *C. wallichii* from Asia, is sometimes available and is a little taller at about 2ft (60cm). The flowers are sky blue, rather small, but carried in large numbers although the overall impact is rather softer than is the case with 'Firmament'.

# D

**Dahlias**

For many years just two seed-raised strains of dahlias were dominant, both of these still stand up well against the competition, and they exemplified the virtues of the best seed-raised types around now. Brilliance and intensity of flowering, ease of propagation, tolerance of less than ideal growing conditions, excellence for cutting, robustness in bad seasons, long flowering period — almost everything you could ask for in a bedding plant. And, of course, like other dahlias they produce a tuber so if you like to, you can save them for the next season. When it comes to the larger types, those intended specifically for cut flowers, showing or for tall displays in the garden, the situation is rather less satisfactory. A year or two ago I grew a selection of standard tall varieties from tubers and the seed-raised equivalents. I started off counting the number of flowers picked from each variety but soon gave up when it became obvious how unproductive the seed-raised ones were. They made tall but sparse plants, flowers were few and the quality of the individual flowers was not only variable but generally poor.

With the small bedding types this is far from the case. They can be treated in just the same way as most half-hardy annuals and they are easier and quicker to raise than many. The seed is fairly large and easy to handle and can be space-sown in trays or large pans. They are happy left until after the usually recommended seed leaf stage before pricking out and once germinated are content with relatively cool temperatures. It pays to prick them out into 3in (7.5cm) pots and some varieties will then be in flower by the time you plant them out after hardening off at the end of May. In the garden they prefer sunshine and a relatively fertile soil, though its basic type does not seem to matter too much. On sandy soil and in dry summers, watering will make a dramatic difference to the display. Slugs of course can be cruel pests, especially shortly after planting out, so take precautions.

If your soil is especially poor, avoid the really dwarf types like 'Figaro' and 'Mignon' which are likely to remain very small and dumpy plants, never producing a very enthusiastic display. One of the most important factors in maintaining a good display, and this is not the first time I have said this, is regular dead-heading. This is especially true late in the season when it is quite common to see a bed of dwarf dahlias covered in nodding, green, rather wet seed heads. They are easy enough to have off, two fingers under the flower, a sharp pull and they come away easily, it really will make a big difference.

The varieties on offer can be grouped into three sections: the dwarf bedders, the taller bedders and the genuinely tall. Two excellent dwarf bedders have arrived from Holland in recent years called 'Rigoletto' and 'Figaro'. 'Rigoletto' was the first, makes about 14in (26cm) and has double and semi-double flowers on bushy little plants in a good range of shining shades. 'Figaro' has appeared more recently and is a couple of inches shorter, more fully double, flowers slightly earlier and is rather more uniform in size — 'Rigoletto' has a tendency to produce some very small plants. And I have seen some 'Figaro' getting on for 2ft (60cm) tall so maybe the seed stock is not being maintained as well as it should be. 'Figaro' does not always appear under that name, some companies list it as 'Rigoletto Improved' or 'Dwarf Rigoletto'; having established the name of 'Rigoletto' with customers, they are loath to change it. The other dwarf one you might find is called 'Mignon', also known as 'Dwarf Dahl-face'. This often makes less than 12in (30cm) in height with small single flowers on rounded bushy plants. The flowers have a tendency to face a little upwards making the impact more positive. These varieties are good in larger containers, small beds and in clumps between out of season shrubs but this group does show a preference for good soil. Look out for the new 'Amoré' too.

Most varieties fall into the tall bedding group where 'Coltness' and the 'Unwins Dwarf Hybrids' have been so popular for so long. The 'Coltness Hybrids' reach about 1½–2ft (45–60cm) with big single flowers and still stand out for sheer brilliance. The last time I had a really close look at the 'Coltness' mixture I found the following colours: white, scarlet, purple with maroon streaks, purple with an orange central ring, yellow with red streaks, apricot with a red ring, pink with white streaks, yellow, pink with a purple ring; nine colours in all. There is a recent new introduction in the 'Coltness' mould called 'Sunburst'. The single flowers are altogether larger at 4in (10cm), the plants are bushy and often reach more than the 2ft (60cm) suggested. There are at least eight colours and the dark green foliage sets them off well. Some of the flowers are so large that the tips of the petals flop.

'Unwins Dwarf Hybrids', and re-selected versions with names like 'Southbank', reach about 2ft (60cm) with mostly double flowers although semi-doubles and even some singles will appear. The colour range is very wide, I gave up after counting ten colours with a good few still to go. There are quilled versions around too which are similar in habit, although generally with more good doubles, but a little more variable in height. These come under such names as 'Dwarf Disco', 'Quilled Satellite' and 'Unwins Ideal Bedding'.

'Redskin' won a Fleuroselect Medal in 1975. It is unique amongst seed-raised dahlias in the dark reddish colour of the foliage but is slightly variable, some plants having greenish tints to the leaves. The flowers are, to be honest, not as impressive as most other varieties, they are smaller and rather less intense in colouring. There was a time when it seemed that this variety had declined in quality. There were fewer, rather dirtier flowers and they appeared smaller but the variety seems to be back to its best now. Feeling that the colour of the foliage was rather oppressive and that the flowering needed a boost, seed companies have introduced mixtures under such names as 'Chi-Chi' and 'Gypsy Dance' which are blends of 'Redskin' and more floriferous dwarf bedding types.

Hursts have worked on an entirely different form of dahlia, the collarette, and this has led to the variety 'Dandy'. The flowers are about 3in (7.5cm) across on plants up to 2ft (60cm) high and the petals are in a good range of pure colours. Around the central eye is a collar of short petals either the same colour as the main flower or contrasting. They are excellent cut flowers and interesting distinctive garden plants. The quality of this variety has varied over the years too, but again seems to be back up to the high standard we expect.

It is when we look at the tall types, those that are analagous to the taller border types grown from tubers, that we see rather a different picture. Simply, they are not worth growing. 'Pompone Mixed' may be 'much more economical than planting tubers' as one catalogue puts it, but if you only get 20 per cent genuine pompoms, is it really so economical? When you also realise that you get fewer flowers of any sort on less neat plants of variable heights. . . A packet of seed will cost you about the same as a tuber. And it is not just the pompom types to which this applies, the giant decorative and cactus types are also rather poor relations and tend to grow taller than their vegetatively-propagated relations

Finally, a new variety due to be announced as a Fleuroselect Silver Medal winner soon. 'Sunny Yellow' was entered in 1985, with a scarlet and a pink relation, and was awarded its medal. Seed not being available in sufficient quantities, the award was held over. It is a short, 15in (38cm), variety and is very bushy with brilliant yellow, almost fully double, but slightly variable, flowers. All are excellent bedding varieties.

**Delphinium (Larkspur)**

Both the annual larkspur and the perennial delphiniums come under this heading although the perennial types are of limited use. The annual larkspur has been introduced to Britain from the Mediterranean for some time and was once a cornfield weed but is now rarely found. It suffers from some confusion over its name. Once known as *D. ajacis* it was then sub-divided into *D. ambiguum* — the larkspur, *D. orientale* — the eastern larkspur and *D. consolida* — the forking larkspur. *D. ajacis* seems to be finding favour again in some quarters and to confuse the issue one specialist catalogue insists they are called *D. grandiflorum* and *D. paniculatum*. To avoid endless tedious botanical analysis, the various varieties will be dealt with by the names under which they are generally listed with their progenitory history largely ignored.

They are hardy annuals, and amongst the most reliable too as they delight in being sown in the autumn and overwintered as well as going in any time in spring from March to May. The colours are rich and pastel — you may think this is an odd combination but if you grow a row of 'Giant Imperials' you will know what I mean.

Most varieties only come in mixtures but there are one or two named ones in separate colours and you will know by now that I will not let them pass by without their being recommended, so here goes. 'Rosamund' has double flowers in bright rose pink and grows to about 3ft (90cm).

This is the one for the border — keep the mixtures for cutting. Try it alongside *Lavatera* 'Silver Cup', *Petunia* 'Birthday Celebration' and a selection of the pinks in the 'Gala' series of geraniums. If you want to put your tongue in your cheek while getting up to all this then put one big orange *Tithonia* 'Torch' right in the middle. 'Ruby King', as you might expect, is similar except for its rich colour. There is also a double white which is very fine and would make a splendid plant in an annual white garden — something which I have never seen but which I will be trying before long.

The dwarf types are becoming more popular and 'Blue Mirror', 'Blue Cloud' and 'Blue Butterfly' all make low rather tangled and twiggy little plants with a long display of flowers in various shades of blue, often with lilac tints. Mixtures in a variety of blue shades from very dark to very pale are on the way. Unlike the other varieties it pays to sow these in warmth in March.

As far as mixtures are concerned they come in various sizes from 'Dwarf Hyacinth Flowered', a double growing to just 12in (30cm). These are excellent for the smaller, more dainty flower arrangements with very dense spikes of flowers in pink and blue, plus some purples. 'Giant Imperial', altogether different in stature reaching 4ft (1.2m), is very aristocratic and a supreme cut flower. The problem with all these varieties is mildew which can rapidly curtail flowering in the autumn once it gets a hold. For cut flowers grow 'Giant Imperials' on a bed system, sowing in rows 9in (23cm) apart and thinning to the same distance. Five rows can go in a bed and the plants supported by horizontal netting. The flowers will be ready for cutting from late June until the end of August and should be cut when just under half the flowers are open. Crown rot can sometimes be a problem causing rotting at the base of the stem; wilting is the first obvious sign but a drench with fungicide will usually solve the problem.

When it comes to the perennials the sad fact is that if sown early to flower in their first summer they really are not very impressive. So you leave them in for another season whereupon they delight you with their impressive spikes and get in the way of next year's carefully planned scheme — gardeners must be adaptable you see.

These are the ones to pick if you want to try the perennials to flower in their first season. 'Blue Fountains' has double flowers in the range of traditional delphinium colours which should stay more or less at its suggested 3ft (90cm) in its first year at least, although when I have grown it recently it has tended to creep up a little beyond that. The 'Connecticut Yankees' are even smaller at around 2ft (60cm) with a good range of shades. 'Blue Heaven' is a selection from 'Blue Fountains' with an especially intense shade and is slightly shorter. 'Dwarf Snow White' looks to be an excellent newish one growing to just 2–2½ft (60–75cm) and although it is not exactly snowy its slightly creamy colour is lovely.

## Dianthus (Pink)

Most bedding pinks are derived from the Chinese pink (*Dianthus chinesis*) a delicate, tufted type with narrow leaves and flowers about 1in (2.5cm) across in red, pink and white shades. This is a perennial plant but one of the first developments many years ago was an annual form 'Heddewigii', the Indian pink, under which you often find the bedding types listed. Most of the best bedding pinks come under this category and there have been some dramatic developments over the years: the flowers have become bigger, the plants bushier and more compact, and they carry more flowers over a longer season. The colours also are more intense, in a wider variety of shades.

'Telstar', which was awarded a Fleuroselect Bronze Medal in 1982 only grows to about 8in (20cm) and although it does not start to flower as early as some other summer bedding plants it continues well into the autumn. The leaves lack the slightly greyish bloom of many pinks and this helps set off the deeper colours in the mixture. There are at least eight colours including a lovely deep red and a white with a large scarlet eye. In my garden they overwintered quite happily and flowered fairly early the following year when cut back in spring as they started to grow. The only problem I had with this strain is that the different colours did not all come into flower together. This is caused by problems in the genetic make-up of the plant but this has been largely solved by the breeders of the 'Princess' strain.

'Princess' flowers in flushes to a certain extent but far less than other varieties. The colours vary in their capacity to keep going and the scarlet and crimson seem the most capable in this respect. They reach about 12in (30cm) with a rather upright habit and the flowers tend to be carried towards the tops of the plants where you can best see them. In the mixture there are quite a few shades including scarlect, crimson, white, red and white bicolour, rose pink, rose and pink bicolour, pale pink, crimson with white edge and more. The flowers are rounded in outline, with slightly fringed edges and are about 1½in (4cm) across. The 'Charm' series of varieties in five colours, two of which have won Fleuroselect medals, are smaller in stature and in flower size and are noticeably intermittent in their flowering habits. You will often find that although you get a brilliant display until mid-August, they will then tend to fizzle out. 'Baby Doll' is different. The plants are similarly small but the flowers are nearly twice as big although this extra attribute really does them few favours. The result is rather gross and unnatural; the flower colours are not impressive — rather odd yucky pinks seems to dominate.

When it comes to carnations from seed most of us think only of the old 'Giant Chabaud' and the 'Enfant de Nice' strains. But the breeders have been busy in this group and produced two outstanding introductions.

The 'Knight Series' is a race of dwarf plants reaching just 12in (30cm). They are very bushy and the colours are exquisite — crimson, scarlet, yellow and white — plus an orange picotee which is pale orange with a dark orange picotee edge and a crimson picotee which is pure white with a fine red edge. All make excellent cut flowers in spite of their modest height and are ideal for bedding and especially tubs and window boxes. Most are well scented. They make good pot plants too.

The other great innovation is 'Scarlet Luminette' which gained a Fleuroselect Bronze Medal. This is taller, at up to 2ft (60cm) and is ideal for cutting as each plant will produce 20–30 flowers. It is fully double, bright scarlet in colour and scented, although not strongly. The stems are stout and stiff and in most areas will not need staking. It is not the earliest plant to flower but will overwinter happily in areas which are not too cold.

In the true perennial pinks there have been some developments too, in particular the lovely 'Lace Mixed' which is a great improvement on 'Loveliness'. This is a form of *Dianthus plumarius,* the pink which is naturalised on walls in many parts of Britain. The 'Lace Series' has good-sized flowers which consist almost entirely of lacy fringes. The scent is strong, they flower in four months from sowing time and are good perennials too.

One variety a little out of the main run is a new dwarf version of the British maiden pink (*Dianthus deltoides*). Called 'Microchip' it reaches just 6in (15cm) but spreads out to over 30cm (1ft). The colours are predominantly purples, lilacs, reds and pink with a lovely white which has a red eye. Indeed most of the flowers will be eyed. It has a light but noticeable scent and will flower in July when treated as a half-hardy annual or in June the following year, on much larger plants, when sown outside in July and transplanted in the autumn.

One of the great advantages of all these varieties is that they do not need too much heat to grow well. In fact if grown too warm they get lank and limp. Sow early, say about February, at a temperature of 65–75°F (18–24°C) but after pricking out keep the temperature down — frost protection will be enough. Only cover the seed lightly and germination should take two weeks at most. Plant out in full sun in soil that is fairly well-drained and, if you like any particular colours in the mixtures of perennial sorts, they can be divided or increased from cuttings.

An extraordinary development has taken place in sweet williams with their conversion to summer-flowering bedding plants. As if there were not enough summer plants without plundering the few spring biennials to create more. However, 'Roundabout' is a very dwarf — 6in (15cm) — plant flowering from June to September in a range of bicolours and spreading almost twice as wide as it grows high. It can even be sown outside and will still flower in its first summer.

Sweet williams are derived from *D. barbatus* from China and Russia. Although there are annual strains, like 'Roundabout', it is as biennials that sweet williams are chiefly grown and this is as it should be. The problem is that they flower a little later than most spring bedding plants and so disrupt the changeover to summer plants. They can be treated as short-lived perennials but tend to deteriorate in the second year so you will wish you had taken them out. The usual biennial treatment suits them very well and some can be left lined out in their nursery beds as they make excellent cut flowers. Once cottage garden favourites, they associate well with lupins and Canterbury bells in a delightful show of early summer colour. 'Auricula Eyed', with a large white eye is a lovely cottagey variety and others with unremarkable names such as 'Superb' are without eyes. The double forms are less elegant.

**Digitalis (Foxglove)**

It may be open to debate as to whether foxgloves have a right to appear at all in a book on annuals and bedding plants but at least one variety, 'Foxy', makes a good stab at being a half-hardy annual and the others can be used as temporary plantings even if they do flower at a time which is a little out of balance with the other biennials. Foxgloves, of course, are amongst the best-loved British wild flowers, and in the wild they reach anything from 9in (23cm) in the scree along the roadsides in north Wales, to 6ft (1.8m) in clearings in West Country woods. Usually purple with darker purple spots ringed in white, pure white or white with purple spots are also found regularly. The foliage of white-flowered plants is paler too. The proportion of white can vary from a single occasional plant to around three-quarters white, which I have recently seen in a number of places in Scotland, but stands of entirely white plants are not common. I also recently came across a purple-leaved form but it remains to be seen how well its seed-raised offspring retain the colour.

Wild foxgloves have the flowers on just one side of the stem which arches slightly and is very elegant. The cultivated forms with flowers all round the stem are different altogether — much more showy but less elegant.

The only variety that can be reliably treated as annual is 'Foxy' which rarely reaches more than 4ft (1.2m) and usually rather less. The flowers come in various reds, pinks, creams and white, all are spotted and they are carried all round the stem. Treated as a half-hardy annual 'Foxy' will flower in five or six months from seed so if it is sown in February it will flower in July. The following year flowering will of course be far more dramatic and also a little earlier. When I tried this, from a March sowing just one spike was in flower by mid-September with five almost out and six more well on the way.

Other types are best used as biennials. The 'Excelsior' strain is the most widely grown and is cheap and reliable. Reaching around 5ft (1.5m) the flowers are densely packed on long spikes and are held almost horizontally so that the markings in the tunnel of the flowers are easily seen. The colour range is especially good and includes purple, pink, primrose yellow and cream, all with impressive spotting.

1 *Antirrhinum* 'Cinderella'

2 *Rudbeckia* 'Goldilocks'

3 *Amaranthus* 'Red Fox' and *Pyrethrum ptarmicaeflorum*

4 *Viola lutea*

5 *Viola* 'Majestic Giants'

6 *Viola tricolor*

7  Swiss chard and *Verbena* 'Blaze'  8  *Chrysanthemum segetum*

9  *Geranium* 'Scarlet Diamond', *Cineraria* 'Silverdust'
   and *Lobelia* 'Sapphire'

10  Marigold 'Inca Orange' and
    Chrysanthemum 'Gold Plate'

11  *Calceolaria* 'Sunshine'

12  *Begonia* 'Non Stop'

13 *Chieranthus* 'Orange Bedder' and *Tulipa* 'Halcro'

14 *Ionopsidium acaule* and *Tulipa griegii*
'Cape Cod'

15 *Geranium* 'Scarlet Diamond'

16 *Centaurea* 'Polka Dot'

17 *Centaurea cyanus*

18  *Ipomaea* 'Heavenly Blue' and *Lavatera* 'Silver Cup'

19  *Cineraria* 'Silverdust', *Cineraria*
'Cirrus' and *Centaurea gymnocarpa*

20  *Ricinus* 'Impala' and *Nicotiana affinis*

The only other varieties of interest are two single colours which are sometimes available — the white form of the wild strain and a lovely apricot. All are of course short lived perennials and soon begin to deteriorate. They are still best raised from seed every year.

Foxgloves show a definite preference for acid soils, partial shade and some organic matter but are generally fairly tough and can usually be persuaded to thrive in most gardens. A little shelter will obviate the necessity to stake the taller varieties. Most of the improved forms look best in informal groups in mixed borders. In wilder situations choose the natural variety and its white counterpart as the 'improved' forms often look incongruous.

**Dimorphotheca (Star of the Veldt)**

This book is dominated by members of the daisy family which seems to include a vast number of useful annuals and dimorphotheca is one of the most dramatic. From South Africa, as you might guess, they can be treated as half-hardy annuals or as hardies as long as you hold off until at least mid-April before sowing. Most of the wild types are orange but in cultivation white, salmon and red are also found. The flowers have an impressive sheen which glistens in the sunshine — the flowers often close in dull weather anyway. 'Glistening White', Highly Commended in an RHS trial in 1982, at about 12in (30cm) is one of the best white annuals and one which seeds itself innocuously about the garden as well. Large patches are rather too stunning but look good in front of *Lavatera* 'Silver Cup' or the purple-leaved castor oil plant *Ricinus* 'Impala'. 'Dwarf Salmon' is more or less as its name suggests. 'Orange Glory' has enormous orange flowers with a black disc which can be softened with blue larkspur and there are a number of mixtures of orange and yellow shades going by such names as 'Aurantiaca Hybrids', Highly Commended in 1982, but some of the softer shades detract from the impact of the more penetrative colours.

**Doronicum (Leopards Bane)**

One of the best and most showy of spring-flowering border perennials there is one variety of doronicum that can be used as a spring bedding plant by treating it as a hardy biennial. The big butter yellow daisies top stems about 15in (38cm) high above clear green foliage, rather shorter than they will be in their second and subsequent years if they are left in. They can either be sown in the open or in a cold frame, transplanted or planted out 6–9in (15–22.5) apart and then moved into their flowering homes at about 9–12in (23–30cm) apart. Grow them with a scarlet or orange tulip. Alternatively, try some bright pansies such as 'Crystal Bowl Red' or 'Azure Blue' in front, or polyanthus. In the late spring they can be given away, moved into a permanent spot or sold off at the WI stall. The danger, if you can call it that, with these perennials grown as spring bedding is that there is always the temptation to leave the plants in place and so the space for a changing display gets smaller and smaller — be ruthless, that is the answer. The variety to pick for the best spring display is *D. caucasicum* 'Magnificum'. The altogether more compact and smaller-flowered *D. grandiflorum* would also be worth trying, especially with pansies.

# E

**Eccremocarp-us (Chilean Glory Flower)**

Eccremocarpus is another much neglected plant, and one of my favourites. It is a vigorous scrambler, clinging by tendrils, with loose spikes of orange flowers, each with a yellow lip. It is not a terribly elegant plant and can be rather brittle but it fairly jumps out of the ground once it gets set. Although usually grown as a half-hardy annual and flowering well in its first year outside, it is in fact a shrubby perennial plant and in most winters in southern Britain only part of the top growth will be killed. In my own garden in Cambridgeshire, in the 1983/4 winter, it was knocked back from 6ft (1.8m) to about 3ft (90cm) and was flowering again in late April. In 1984/5 it was cut to ground level but was shooting from below ground in early May and by the autumn it had reached 5ft (1.2m). Although usually treated as a half-hardy annual, there are other ways of treating it. Often, self-sown seedlings appear under the plants and these can be moved to new sites or left to furnish the lower areas of plants which have survived a number of winters and become rather bare.

Another way, for those in colder areas who keep a frost-free greenhouse during the winter, is to sow in August, prick out into 3in (7.5cm) pots, move to 5in (12.5cm) pots in the autumn and keep them frost-free for the winter. By the time you come to plant them out in May after a hardening off period they will already be in flower.

As well as the type described there is a golden-flowered version and a rather dull pink one. Ralph Gould, who has bred so many wonderful annuals, has recently come up with a mixture of colours known as the 'Anglia Hybrids' with flowers in pink, scarlet, crimson, yellow and orange. I grow the standard orange version on a sunny fence in fairly well-drained soil with *Ipomaea* 'Heavenly Blue' set at the bottom to make a lovely combination as they twine together. It might be worth running one up a summer-flowering ceanothus such as 'Gloire de Versailles' and the golden one could be mixed with *Cobaea scandens* which likes much the same conditions. But if you live in a cold area and are tempted to grow it in a cold greenhouse you will get a lovely canopy of foliage but the flowers will all be pressing on the glass and you will not be able to see them.

## Echium (Viper's Bugloss)

A dramatic British plant, echium is most frequently found on grassy dunes and is very impressive on The Gower in South Wales where it stands up, big and blue, in the closely rabbit-cropped turf. The common name is interesting. A number of plants had the name 'bugloss' which is derived from the Greek meaning ox-tongue — the original Greek word applied to anchusa. The 'viper' comes in because it was thought that the seeds resembled snake's heads and Dioscorides recommended it as a preventive and cure for snake bites. It is a biennial, like so many in the borage family (which includes forget-me-nots as well as anchusa) but is usually grown as an annual. Sown in the autumn it makes flat, rough rosettes and throws up the flowers from June onwards. The wild species grows to about 3ft (90cm) when at its best and you can get the straight species from some companies. The flowers are exquisite in colour — an intense blue unlike almost any other.

The cultivated forms are all a great deal shorter. 'Blue Bedder' is only 12in (30cm) high and the various mixtures are about the same height. The mixtures are generally in rather soft shades — pinks and mauves plus white, many with darker streaks. All those available are good in sunny sites and dry soils — this is a group where poor soil really does make a difference to the quality of the display. Also, sowing in the autumn pays off more often than with many others as they are natural biennials. The cultivated forms are sometimes listed under *E. lycopsis* or *E. plantagineum,* alternative names for the purple viper's bugloss which is native to the extreme south-west of Britain. This is not a plant to put near the children's sand pit as bees seem to be especially fond of it and this could be a hazard; sometimes the plants are swarming with them.

**Emilia (Tassel Flower)**

Also known, for reasons which escape me, as Flora's Paintbrush this pretty little annual is not a real stunner but its colour is unusual and it is a great deal more delicate than most half-hardy types. It grows to about 18in (45cm), maybe a little more, with tall, stiff but dainty waving stems topped in heads rather like loose clusters of ageratum flowers — except for the colour. *Emilia coccinea* as you might expect has red flowers, a slightly orangey red, although you sometimes find it in a mixture of scarlet, orange and gold. The usual half-hardy treatment suits it well although it can be sown outside in late April where it is to flower. It flowers right through the summer and makes pretty, long-lasting cut flowers. This is another botanical wanderer and may be found in catalogues as *E. jauanica, E. sagittata* or *Cacalia coccinea*.

**Eschscholtzia (Californian Poppy)**

As with so many plants, the wild situation tells you what it likes in the garden. From California, this almost unspellable little annual likes the maximum of sun and the minimum of boggy soil — one of the few for whom the old dictum of poor sandy soil really does apply. There are two species generally available in the UK, *E. caespitosum* and *E. californicum*. As the name suggests *E. caespitosum* is an annual alpine (*caespitosa* means growing in tufts). It reaches just 6in (15cm) and in the form most usually available, called 'Sundew', carries enormous numbers of lemony flowers each with a strong scent. 'Miniature Primrose' is similar. This is one of the few annuals which alpine purists could surely admit to the rock garden. Mind you, some prejudices die hard.

The Californian poppy is rather taller at about 12in (30cm) and larger in other respects too — not least the impression it makes on the garden scene. The foliage is fine and has a tendency to sea green shades and the flowers come in sparkling mixtures or single shades. 'Cherry Ripe', 'Orange King', 'Milky-White' and 'Purple-Violet' are self-descriptive varieties and there are a number of single, semi-double and double mixtures but only the singles have the real intensity that typifies the group. 'Monarch Art Shades' is a good double and semi-double mixture but in single mixtures the stocks vary greatly. The best have rich purples as well as hot oranges. The flowers on all of them are of impressive size — about 3in (7.5cm) — and are followed by long curved seed pods. Although these are quite interesting it pays to remove them to make sure the flowers keep coming.

This is another plant for fiery combinations. The white is less satisfactory but the purple is good and the orange and yellow supreme. They are good scattered thinly along the edges of gravel drives — as long as you remember to leave the patches untreated with the weed preventer that will keep them from engulfing the whole driveway. One tip to remember: however much you might be tempted to treat them like half-hardies, resist. They do not fit well into the normal transplanted bedding scheme and they do not transplant well anyway so always sow them where they are to flower.

**Eucalyptus**

These are astonishing foliage plants for summer bedding, making trees of silvery leaves that associate well with a wide variety of annuals and which may survive for two or three winters making small trees very rapidly.

*E. globulus* is the one usually grown for summer bedding as it germinates easily, grows quickly and is particularly well-coloured. You may come across a dwarf variety, 'Compacta', but this is not suited to bedding being too slow growing. Sow in March at 70°F (21°C) and prick out into peat pots as early as possible. If left in the seed pot too long they may not take kindly to pricking out. Grow on fairly warm with other half-hardies and when the roots start to penetrate the walls of the peat pot move into 5in (12.5cm) pots. Keep them growing and harden off well before planting out. The trick is to get them as large as possible when planting out without their being in the least pot-bound. They are temperamental plants and do not always take to planting out, especially from a pot full of roots.

It is no bad thing to put a 3ft (90cm) cane in to support the plant when doing the final potting. This cane will also serve to support the plant after planting. Tie the stems in only loosely so that after planting the cane can be pushed in another 6in (15cm) for extra stability. Take care not to poke your eye on the canes in the greenhouse — some gardeners upend a 2½in (6.5cm) plastic pot on the cane as a precaution but it does not look very elegant. Water thoroughly with a liquid feed before and after planting. They seem happy in most soils except those which are wet and although they do best in the sun will make good plants in the shade of walls and fences though not under trees. Do not pinch the tops out unless you are unable to give the plants plenty of space to grow — they will branch naturally and reach greater height if left unpinched. Only if there is insufficient greenhouse space to spread them out should they be pinched at 6in (15cm).

In warmer areas they will make 5–6ft (1.5–1.8m) in their first summer so you do not need many plants. They will be killed by the average winter but in mild seasons and in sheltered spots much of the more substantial growth will survive. As I write I look out on one whose top growth was killed in the 1984/5 winter. It was two years old, making a well-branched 6ft (1.8m) specimen in its first year in a spot where it had the sun for about a third of the day. The 1983/4 winter killed all the tips but left everything else undamaged. I cut it back to about 4ft (1.2m) in late April and by the autumn it had reached about 8½ft (2.6m), elbowing a few neighbouring plants aside in the process. In its first season it had as neighbours *Chrysanthemum foeniculaceum*, *Petunia* 'Chiffon Magic', *Salvia* 'Victoria' and some *Polygonum bistortum* 'Superbum'. The white sweet pea 'Diamond Wedding' set to ramble through the whole lot suffered from the root competition and never quite made it. In the second summer, as the top appeared out of the shadow of the fence into day-long sunshine, *Ipomaea* 'Heavenly Blue' twined around its branches making one of my all time favourite associations. It looks well in all-white arrangements too with *Lavatera* 'Mont Blanc', *Petunia* 'Recoverer White', the chrysanthemum again and the white *Antirrhinum* 'Giant Forerunner'. One more suggestion for something really outrageous, if you have a plant that has overwintered in a really sunny spot, is to put *Ricinus* 'Impala' with its glistening almost maroon foliage like giant sycamore leaves next to the eucalyptus. *Tithonia* 'Torch' in bright orange might make a stunning trio.

**Euphorbia (Spurge)**

An annual and a biennial make up the contribution to bedding in this apparently heterogeneous group, which includes border perennials, enormous succulents, irritating weeds and one of the most ugly of all plants which the sensitive gardener so far forgets himself as to grow — the poinsettia.

Sometimes called snow on the mountain, a little fanciful I feel, *E. marginata* is an easy hardy annual, making single stems branching mostly at the top. The foliage is mid-green but the upper foliage and the bracts around the insignificant flowers are, to a greater or lesser degree, margined in white. In some bracts the green part is reduced to a pale yellowish line along the middle. Once established, this plant tends to sow itself about the garden in a fairly unobtrusive way but you must be careful when removing unwanted seedlings. Like all euphorbias, the milky sap which exudes from cut and damaged stems and leaves is highly irritating and should be kept out of eyes, mouth and wounds and off the skin too.

This plant is one of the very few where the variegation is genetically controlled — so the plants always come true from seed. Most variegated plants can only be raised from cuttings or by grafting. The plant reaches about 2ft (60cm) high and should be thinned to about 12in (30cm). There is also a rather shorter variety called 'Summer Icicle'.

The caper spurge or mole plant is a swindle. Every now and again someone who should know better recommends this as an ideal solution to the mole problem and as sure as a hangover follows a heavy night with a friendly bottle, moles will be upending their plants all over the country in no time at all. Occasionally, as I myself have woken up clear headed after a jolly evening, moles have been known to vanish from gardens where caper spurge has been planted. But I think this comes under the heading of 'the exception that proves the rule' — whatever that means. But this is nevertheless a plant which is worth growing. It is native, or possibly naturalised, in the southern half of Britain where it generally grows in woods. It is common in southern Europe and was once cultivated for its seeds. It grows 2–3ft (60–90cm) high and is usually unbranched. The leaves are long and narrow and stand out at right angles to the stems in a tightly symmetrical pattern. The flowers are the usual spurgey olive. It is a biennial but once you have it you will never get rid of it and severe thinning and removal to the compost heap is often necessary.

**Eustoma (Prairie Gentian)**

This is an annual member of the gentian family recently given some publicity under its old name of *Lisianthus russellianus*. It grows 2–3ft (60–90cm) high, has upright, rather lithe stems with grey green leaves and the flowers, which are bell-shaped, come in a number of delicate colours — violet, pink, cream with a yellow centre. The original species has purple flowers with a black centre while the newly named and newly released *E. grandiflorus* 'F1 Mixed' has that greater variety. A double-flowered form, 'Prima Donna', has recently been introduced. They make superb cut flowers, lasting up to four weeks, but unfortunately they need a very long growing season and some special treatment. Start in January and after sowing put the pots in polythene bags at around 70°F (20°C) to keep the moisture level high. Pot them up at the three leaf stage, water carefully to avoid encouraging damping off and plant them out when danger of the last frost is over. They should start to flower by the end of July. Alternatively, they can be sown in the autumn and overwintered.

# F

**Felicia (Kingfisher Daisy)**

With so few really good blue summer annuals, it is a mystery why these splendid, penetrating little plants are not used more often. Penetrating, I might say, only in their intense blue colouring; they could hardly be described as invasive. They come from South Africa and only two are in cultivation generally. *F. amelloides* is in fact a perennial but sadly not hardy. It makes a more or less rounded plant which can be up to 2ft (60cm) high but it pays to pinch out the shoots early on to keep the plant fairly compact. The flowers are 1½in (4cm) across, a delightful sky-blue shade and once they have started to appear they go on and on. Plants can be lifted in autumn or cuttings taken in late summer and overwintered in a frost-free greenhouse or even on a windowsill where they will often flower till spring and beyond. It is sometimes possible to find plants of a very large flowered variety, 'Monstrosa', but these must be kept going by cuttings if the plants are to retain their large flowers. If growing from seed, treat them as half-hardy annuals planting out after the last frost.

*F. bergeriana* is much smaller, only reaching about 6in (15cm) and far more spreading. The flowers which are smaller too at ¾in (2cm), are even more striking being a vibrant kingfisher blue in colour. This is an ideal window box and hanging basket plant but should not be expected to overwinter as it is strictly an annual. This is another one of those choice few suitable for planting in gaps on rock gardens and raised beds. Unlike *F. amelloides* which has bright, shining green leaves, this species has slightly grey foliage. Both these species are best in full sun and preferably in a gritty, well-drained soil. In the south-west, *F. amelloides* might well last the winter outside. Grow it with darker blues like *Salvia farinacea* 'Victoria' or with a pale primrose such as *Petunia* 'Brass Band'.

# G

**Gaillardia (Blanket Flower)**

One of the least trendy of annuals, maybe suffering from the profusion of daisy flowers that can be grown as summer annuals in addition to the vast number of perennials in our borders. Nevertheless, they make bountiful border plants flowering for long periods often in fiery tones. Gaillardias are good cut flowers too, lasting as long as almost any other flowers. There are just two grown widely — an annual and a perennial, although the perennial is perfectly happy treated as an annual. *G. aristata*, sometimes listed as *G. grandiflora*, is a perennial from North America which arrived in Britain in the early years of the nineteenth century. In catalogues it is represented by a number of varieties. 'Burgundy' grows to about 2ft (60cm) and is an unusual shade of deep wine, 'Goblin' is half the size and has brilliant yellow flowers with a ragged red inner zone around a red eye. There is another type which comes under such headings as 'Re-selected Hybrids', 'Large Flowered Hybrids' and the like which are mainly semi-double with hot yellow, crimson and gold shades with bicolours. They can all be sown outside in June or July or in the greenhouse in early spring and treated as half-hardy annuals.

The other type grown is a true annual *G. pulchella* and is usually available as a fully double strain. 'Double Mixed' has fully double flowers in old gold and various coppery shades, 'Lollipops' has fully double flowers too, in a slightly wider range of shades including cream and there are also some bicolours. The flowers are up to 3in (7.5cm) across so are quite dramatic. Again, sun and a soil that is not too claggy is ideal but the taller types will need staking as the heads seem to get easily weighed down by the rain.

**Gazania**

Years ago named varieties of gazania were widely used in parks and large private gardens and these were raised from cuttings every year, the plants overwintered in cold frames. Many of the varieties had almost white, divided leaves. The flowers were big, yellow or orange daisies, the plants usually rather dwarf and the flowers had the infuriating habit of closing up whenever the weather was less than perfect. Hardly a plant for the British climate. But in recent years breeding work has much improved the plant from the gardener's point of view although most of the varieties have green rather than silver foliage. The flowers now stay open rather longer in less than ideal conditions. In gazanias the opening and closing of the flowers is controlled partly by light intensity and partly by temperature. So if flowers are cut and brought indoors where it is constantly warm they will open well; outside some degree of sunshine is usually needed. The amount of sun required is steadily being reduced and doubtless before long a mutation will occur which will lead to all day — and all night — gazanias, but not yet.

The silver-foliaged seed-raised gazania has just been introduced, it is called 'Carnival' but the foliage is better described as silvery green rather than silver. There are only three other strains which should be considered at present: 'Chansonette', 'Mini-Star' and 'Sundance'. 'Chansonette' was the first and comes in mainly bronze, orange and yellow shades plus some with a slightly pinkish tinge. The leaves are green with a silvery underside and the plants reach about 8–9in (20–23cm). The 'Mini-Star' range, of which both the tangerine and the yellow have been awarded Fleuroselect Bronze Medals, is about the same height and the foliage is dark green and glossy. The mixture of shades is rather less generally available than the two award-winning colours but, as well as the usual fiery shades, it includes white, beige and a good pink. Rather taller at 12in (30cm) or slightly more, the 'Sundance' range also has exceptionally large flowers — some I have grown have been fully 5in (12.5cm) across. The colours do not include the pinks and whites but concentrate on the darker crimson, orange, rust, bronze and mahogany shades and recently yellow and red striped forms have been added to the mixture. The long stems make this ideal for flower arranging and in the garden it seems to have the best tolerance so far of dull and cool conditions. Even in the soggy summer of 1985 this variety did pretty well. 'Carnival' is about the same height and includes various pink shades as well as the usual orange, red and rusty shades; there are some lovely striped forms too.

Give them all the standard half-hardy annual treatment but of course, like most tender perennials, you can pick out the colours you especially like in the mixture, take a few cuttings in the autumn in a sandy compost and overwinter them in a cold frame. As you might expect they do best in an open position that gets the best of the sun. They germinate happily in fairly cool conditions, 60°F (15°C), and can be planted out in early May if properly hardened off first; they are a little hardier than most half-hardies. Grow them with other sun lovers — the open habit of 'Sundance' makes it especially suitable for filling gaps in mixed borders.

## Geranium

There has been a revolution in geranium growing these last few years. Whereas all geraniums used to be raised from cuttings and plants overwintered from year to year, they are now commonly raised from seed. And as far as the home gardener is concerned there still seems to be some doubt as to whether the seed-raised varieties really are better. The main reasons for the old cuttings-raised varieties becoming less popular were degeneration of the stocks, the difficulty and cost of overwintering, the susceptibility to diseases like virus and blackleg and the publicity which surrounded new seed-raised strains.

The increasing degeneration of the old varieties was caused by virus diseases, in particular one called pelargonium leaf curl virus. This is an unfortunate name as the leaf curling symptoms are minimal. Pale yellowish spots are more often in evidence and they get darker and eventually lead to dead patches; the leaves then become puckered. An overall debilitation is noticeable and flower production becomes less prolific. The virus, together with one or two others, is soil-borne but the precise carrier is not known. Varieties like the old favourite 'Paul Crampel' became so infected that most people gave up growing it. There has been a lot of scientific work done in this field and quite a few varieties are available which have been completely freed from virus. The problem is that only a few nurseries have the special stock.

Overwintering can still be a problem. Modern fungicides certainly help and the use of greenhouse insulation and thermal screens will reduce the cost substantially. Unfortunately the best method remains to keep them growing slowly in a cool greenhouse but the cost of this can be prohibitive even if only one plant of each variety is kept.

The publicity that surrounded the new seed-raised types obviously did the old varieties down and there were good reasons for growing seed-raised types, especially for nurseries. The problem of overwintering was eliminated and with it many disease problems, especially virus. Nurseries could also fit geraniums into a production schedule for other seed-raised bedding plants. For the home gardener the virus problem was solved but new problems appeared. High temperatures were needed for good germination and flowering was late. By the time the plants flowered they were big, leafy and not too elegant. Again, nurserymen had an answer in a dwarfing chemical, Cycocel, which gave a dwarfer and an earlier flowering plant, but this chemical is not now available to home gardeners. Modern varieties flower much earlier and on smaller plants so this problem has been partially overcome at least.

Recently I tried a combination of the old and new systems. In the autumn I dug up one plant of 'Video Scarlet', a dwarf seed-raised variety, and kept it warm on the kitchen windowsill all winter. It always had at least two heads of flowers. In the spring I got five cuttings from it and they all rooted. By planting out time the parent had quite a few flowers, the rooted plants had at least one head but the seedlings of the same variety were nowhere near producing buds. It was weeks before any of the seed-raised plants matched the flowering of the cuttings-raised 'Video Scarlet'. There are conclusions to be drawn from this. If you heat part of your greenhouse it pays to overwinter. Section off a small area to keep costs down.

It is a good idea, with this in mind, to take some cuttings of your favourite plants. Geraniums root very easily from cuttings. They will root at almost any time of year as long as they are fairly warm but the winter is obviously unsuitable for the home gardener. To have sturdy young plants for overwintering take cuttings in August and early September. You will be able to capitalise on these overwintered plants by taking cuttings from them in late March although it depends, of course, on the season and the temperature you can maintain. Cuttings will root if they are 6in (15cm) long or simply made up of a short piece of stem with a leaf attached, but cuttings about 3in (7.5cm) long are ideal. They should be of non-flowering tips but flowers can be removed if you are short of material; lower portions of stem can be used too, although the resulting plants may be less elegant. The standard seed compost described on page 17 is fine and you can either root them singly in 3in (7.5cm) pots or five or six to a 5in (12.5cm) half pot. The chances are that they will all root so if you want to save a job, and have the space, set them singly — although of course you will need to use a potting compost rather than a seed or cuttings compost to give them enough nutrients. There is no need to use a hormone rooting powder as they will usually root very easily and they do not like too much humidity as the slightly hairy leaves may rot. As long as they are shaded from strong sun they will root without too much trouble.

Research has revealed that the most economical way to raise geraniums from seed is to sow early and grow cool. By sowing at the beginning of October and after pricking out and growing on at 45°F (7°C), they will flower in the last week of May. This saves you about 35 per cent of the heating costs of the old system of 65°F (18°C) right through starting in early January, although flowering is a couple of weeks later. By starting in January and growing at the reduced temperature you save only a little more fuel but they will not flower until mid-June. I suggest the following all-round programme.

(1) Take cuttings in August or early September and grow on in a cold greenhouse, setting a minimum temperature of 45°F (7°C) once the heating comes on.
(2) Order seed of new varieties as soon as the catalogues arrive.
(3) Sow seed in a propagator at 75°F (24°C) as soon as it arrives. Drop to 65°F (18°C) when it germinates and down to 45°F (7°C) after pricking out.
(4) Order new plants for spring delivery.
(5) Spray plants against botrytis regularly with

           propiconazole, benomyl or bupirimate and triforine
           (or better still ring the changes).

(6) The time to take more cuttings depends on the greenhouse temperature but the end of March is usually about right.

(7) Pot up new plants as soon as they arrive and pot up rooted cuttings promptly too.

(8) Harden them all off from the beginning of May and plant out at the end of May when, depending on eventual sowing dates of seed-raised types, all should be flowering.

Another thing I have looked at in relation to seed-raised varieties is the best way to raise the seed. I tried three methods:

(a) Conventional sowing and pricking out at the seed leaf stage into trays.

(b) Sowing a single seed in Jiffy 7 net pots and potting up when the roots penetrated the net.

(c) Sowing singly in Jiffy Strips (peat pots about ¾in (2cm) square which come in double strips of 12) and potting up when the roots penetrated the sides.

The same variety was used for all and they all survived well, all made good plants and all flowered together. So it really does not matter which you use. Recent research has shown that seeds can be moved direct from the seed tray to the final pot with a hold up of only a few days in first flowering day and this may be more convenient.

Seed-raised varieties are now so good and so varied that whatever your taste you should be able to find some you like. Unless you like doubles. In spite of a lot of work in America, the double-flowered varieties around now are very late to flower, make large leafy plants and the flowers easily succumb to botrytis. Of the others, the following are ones I have found suitable for the home gardener.

*'Gala' Series.* A range of nine British-bred varieties which are naturally shorter and bushier in the earlier stages than most other types. The varieties available in the series are:

'Amaretto' — Deep salmon with strong leaf zoning.
'Coral' — Deep salmon with copper undertones and strong zoning.

'Flamingo' — Pale salmon with no zoning.

'Lipstick' — Delicate coral pink with pale leaves.

'Highlight' — Light magenta with a red eye and plain leaves.

'Redhead' — Cerise red with strong zoning.

'Rose' — Rose pink with a white eye and good zoning.

'Sunbird' — Brilliant scarlet with good leaf zoning.

'White' — Probably the best of all whites, but of course no zoning.

The garden performance of this series is excellent and their relatively dwarf habit makes them ideal for the home gardener. Not all the colours in the range are available separately but more are likely to find their way into catalogues and the mixture will give a fine show.

*'Video' Series.* From the same British breeder as the 'Gala' series, the 'Videos' are even more dwarf and quite distinct from other types. They make little more than 6–7in (16–18cm) with unusually dark foliage. There are just four colours at present, scarlet, salmon, rose and blushed white though more are promised. In the garden they look well in containers of all sorts, including baskets, and are ideal for small clumps and beds. They make excellent pot plants too, eventually making broad plants covered in flowers.

*'Diamond' Series.* Probably the best of all for performance in the home garden, in more austere conditions they are less good. The so called 'Multiflora' habit is quite different from that of most others with low spreading growth throwing up vast quantities of flower heads from low down. I have had 22 heads open on a plant of 'Scarlet Diamond' 15in (38cm) across in September. 'Scarlet Diamond' was the second Silver Medal winner in Fleuroselect trials, 'Cherry Diamond' won a Bronze in the same year. 'Rose Diamond' has now appeared and there is a white variety too.

The scarlet is undoubtedly the best with brilliant flowers almost hiding the foliage; the cherry is slightly variable in colour and a little loose in habit by comparison. The rose variety is a lilac rose pink with white eye on the upper two petals. It is slightly more compact than the others and a little less floriferous. The white is noticeably less good and rarely seen. The 'Diamonds' are the earliest varieties to flower of all the ones I have grown but in some gardens there can be a problem at the other end of the season. On poor soils and in hot, dry seasons they tend to lose their lower leaves but still try to flower prolifically. The result is that the plants burn themselves out during August. This is often apparent on trial grounds where seed companies like to treat their plants hard so they can assess how they put up with difficult conditions. In gardens, we look after our plants rather better and under those conditions the show will be unparalleled.

*'Breakaway' Series.* Yet another variety from Floranova, the British flower and vegetable breeders: it shows how working in British conditions produces varieties which thrive in Britain. The 'Breakaways' are the better of the two strains of so-called trailing geraniums. They are not ivy-leaved types, they are zonal pelargoniums with a horizontal habit of growth. When planted in a basket they will not hang down, but will arch gently. There are three colours, which are sometimes available separately, coral red, soft salmon and red, although the red is slightly variable. Three plants will fill a 10in (25cm) basket and they also do well planted in the border. They are less dramatic than other varieties when planted out but can spread to more than 2ft (60cm) across and their more restrained, but still very colourful, display makes them ideal for spaces in a mixed border where really intense colours may detract from the other plants. 'Red Fountain' (also known as 'Orange Cascade', which tells you how variable the colour is!) is similar but less impressive although it is available in a wider range of colours.

*'Lucky Break'.* This is a new mixture of colours that is different from all the others recommended so far in that it is not an F1 hybrid. Most geraniums which are not F1s are a waste of money in that they flower later, the plants grow very big and leafy and the number of flowers they carry is very few. On the day when my 'Scarlet Diamond' had 22 flowers and 'Cherry Diamond' had 18, none of the plants in the 'Fleuriste' mixture had more than three; even 'Sprinter' had 12. But 'Lucky Break' is different. It is very much in the 'Diamond' mould although it originates at Asmer in Britain rather than in Germany, and even suffers the same burn out in inhospitable conditions. The plants flower very early. Some gardeners had them in flower before the 'Diamonds' in 1985, and the plants stay fairly small throwing up lots of flowers from the base. The plants also vary rather in size;

the flowering is a little less impressive than the 'Diamonds' but is far better than any of the other F2s and quite as good as many F1s. But the great thing about 'Lucky Break' is the price. At the very least seeds should be one-third of the price of F1s if a little more expensive than other F2s. The raisers of 'Lucky Break' sell it to the trade (they do not deal with home gardeners) at a quarter of the price of 'Sprinter Mixed' or of the standard F1 mixtures. The colours in the mixture are scarlet, crimson, glowing red, salmon, pink and appleblossom. There is a lilac on the way too. You will also find a variety called 'Dame Fortune' listed, as you might guess from the similarity of names, it is related to 'Lucky Break'. In fact it is 'Lucky Break' plus a white which gives a lighter mixture; 'Lucky Break' is heavy on reds.

Apart from these series and mixtures there are of course many other varieties in a wide range of colours. The only types that are not available from seed are those grown for their coloured foliage which are so popular as cuttings-raised plants. What you grow is partly a matter of which colours you like, which varieties your favourite seed company lists and the ones towards which you just have a positive feeling. Of all the others, these are some that I have grown and found provide a good display.

'Solo' is a classic red geranium with good zoning and a brilliant display that beats the rest of its type. 'Grenadier' is another good red with excellent zoning and was the standard variety outside Buckingham Palace in London for some years; it is also known as 'Jackpot'. 'Sprinter', another red, is still probably the best known of all geraniums and although it flowers pretty well in the end, it is less good early on and eventually gets rather large.

'Picasso' is a dramatic magenta and very piercing; this is a shade some people love and some hate but most will at least admit that the zoning is good. 'Cherie' is about the most dramatically zoned of all seed-raised types; it is very strong, distinct and clearly defined. The flowers are salmon and just the right shade. The best in the very pale appleblossom shade is 'Appleblossom Orbit' with lovely soft flowers and nicely zoned leaves — a delightful, delicate colour and indispensable. Oranges are few and far between but 'Sundance' is excellent and a very prolific plant with intense flowers.

There is quite a restricted range of white-eyed types, and 'Hollywood Star', after a lot of publicity, is the current leader although the heads do not carry an enormous number of flowers and the effect is rather thin. The white eyes are quite large giving a very pale appearance, while 'Bright Eyes' has smaller and more distinct eyes; take your choice. A forthcoming British introduction is 'Eyes Right' which is an eyed type, but the flowers are pink or rose with a darker scarlet or cerise eye. This is the first distinctly dark-eyed type and looks to be excellent but seed production problems have delayed its introduction.

Recently a new series has arrived on the market, the 'Pintos', also known as the 'Pulsars'. These are 'Multiflora' types in red, rose and salmon with good zoning and lasting especially well late in the season.

Finally, there is the first ivy-leaved type to be available from seed, 'Summer Showers'. In spite of being a Fleuroselect Silver Medal winner in 1985, this is a very disappointing variety. The publicity campaign has swept into action, billing it as the best new variety for years but unfortunately the actual plants do not live up to the publicity. It is quite a breakthrough from a technical point of view but horticulturally it is not a success. When compared with the existing vegetatively-propagated named varieties, well, there is no comparison. 'Summer Showers' produces fewer flower heads, with fewer flowers in each head and the individual heads are very open and lax. It makes a long trailing plant but because the internodes (stems between leaf joints) are so long the flowers are too well spaced out. The colour range is quite good, and includes a rich purplish shade but, in spite of what the catalogues say, I have yet to see a true white. The worst thing about 'Summer Showers' is the price — it will cost you two or three times as much as other geranium seed — and at that price it really is better value to buy young plants of an established named variety. The problem with this variety, like the seed-raised strawberry 'Sweetheart', promoted as a hanging basket plant, is that it has been released too early. The American breeder who developed 'Summer Showers' has spent many years on the project and there is obviously commercial pressure to try and bring in some financial return. A few more years and a far better variety will doubtless emerge, and it may be that 'Summer Showers' will get better but in its first year of availability to home gardeners it is probably best to stick to the old varieties. How it came to be awarded a Fleuroselect Silver Medal defies comprehension.

Delicate annuals with finely cut foliage, leptosiphon and linanthus are sometimes included under gilia but catalogues tend to list them separately. They are long-flowering plants which are easy to raise given the usual hardy annual treatment. *G. capitata* is the taller of the two usually found and can reach 18–24in (45–60cm). It has rounded heads of soft blue flowers on slender stems and is good for cutting or for the more informal border. *G. tricolor* is quite different: rather than the soft pin-cushion-like flowers of *G. capitata*, it has open flowers with colours in rings — yellow in the centre, then a ring of maroon, then green and an edge of soft blue or purple. The foliage is rather like that of mayweed and the plant reaches about 12–18in (30–45cm). Again it is a border filler rather than a true bedder but is long flowering and has great charm for the less organised garden.

**Gilia (Birds' Eyes)**

The yellow horned poppy is the one usually cultivated and is one of the most striking of British native plants; many holiday makers will be familiar with its big butter yellow flowers and silvery grey leaves. It grows on shingle banks, usually not far above the high water mark all around the British coast except in the far north. This will tell you the kind of soil it needs — clay and excessive water retention are not conditions helpful to a dramatic display. In fact the plant is a perennial but can be sown early in spring and will then flower in its first summer. It does not produce a great profusion of flowers so it is important that the soil be not overrich otherwise foliage will predominate. Mind you, the foliage is very attractive and it is worth growing for that alone. After flowering another feature, interesting rather than colourful, will be seen. The seed pods are up to 12in (30cm) long and grow in a gentle curve making them quite elegant.

**Glaucium (Horned Poppy)**

With clarkia, godetia is another of the basic hardy annuals and one of the few I remember from childhood days when Mum, usually without a great deal of success, tried to foster some semblance of horticultural enthusiasm in her football fanatic of a son. In spite of this early disregard, godetia now appear in my garden every year; they are easy to grow, not too expensive, rarely suffer from ailments, come in a range of soft shades and are good for cutting. *G.*

**Godetia**

149

*grandiflora* is the wild species from which most named varieties derive and this is a native of California, like so many annuals. Whether the double flowers are more, or less, elegant than the singles is a matter for some argument. Personally, as you might expect, I grow singles rather than doubles but all are good garden plants. All have the additional advantage of modest height and the lack of any need for staking, except in very windy sites and of course they last well in water and are especially useful in what you might call cottage-style arrangements — a development of the 'haphazard' school of flower arranging. In the garden the tones are sufficiently subtle and distinct to be used in carefully planned schemes. A book published in the 1950s listed 33 different varieties, but things have changed. Now you will not find more than about nine.

'Sybil Sherwood' is probably the best known. Not sure who 'Sybil Sherwood' was (or is) but her flower is salmon pink grading into white at the edges; a lovely cut flower. 'Duchess of Albany' is pure white, slightly frilly and is a good bedder with silver *Cineraria* 'Silverdust'. There are tall and short double mixtures and a single mixture which is a little too strong on the magenta shades to sit happily in many situations.

## Grasses

Apart from the lawn, and maybe some pampas, many gardens have no other grasses at all. And that is a great shame because not only are they easy to grow, they add quite a different atmosphere to the garden. The manner of growth and the flower forms are so distinct and appealing that no gardens, and especially those looked after by keen flower arrangers, should be without them.

*Avena sterilis* (Animated Oats) — A bizarre common name for a very curious plant. It reaches 2–3ft (60–90cm) and apart from being a good dried flower with its oaty heads, it has a very strange habit. When the ripe seeds are removed from the plant and set down they will move about apparently of their own volition. What happens is that variations in humidity cause expansion and contraction of certain parts of the seeds and this causes movement. Interesting but not much more.

*Briza* (Quaking Grass) — Two species are grown: the greater (*B. maxima*) and the lesser (*B. minor*). Both have

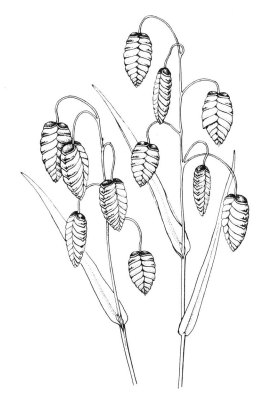

*Figure 7.5* Briza
maxima

the same tapering heads of papery bracts hanging delicately from slender stems. They make an interesting rustly noise in the wind. One reaches 18in (45cm) and the other 12in (30cm) with the flower heads reduced in size, in proportion — they are about 1in (2.5cm) long on the larger version. Both are pretty in the garden and in dried arrangements and like a dryish soil in sun. You will sometimes come across the smaller version growing wild in the south of Britain in dryish arable fields — but do not pick any heads, it is getting rare and you can buy seed. Both are hardy plants which can be sown in spring or autumn. If you need a lot for drying, sow a row on the allotment.

*Coix lacryma-jobi* (Job's Tears) — Arching stems about 2ft (60cm) high carrying relatively large, globular seeds about ¼in (6mm) across which are used as beads to make bracelets and necklaces. An interesting but hardly spectacular plant.

151

*Hordeum jubatum* (Squirrel-tail Grass) — The common name sums up this typical member of the barley family fairly well. It grows to about 18in (45cm) with silvery green, or sometimes brown, nodding spikes of silky heads. The usual hardy annual treatment applies. It is elegant and pretty in the garden and excellent dried.

*Lagurus ovatus* (Hare's Tail) — An upright grass with beautifully soft oval heads that look and feel like a kitten's paws. It grows to about 18in (45cm) and branches very well from low down giving a mass of stiff, though not coarse, stems which are lovely in dried arrangements. It is less hardy than some of the other grasses and is best sown outside in April. One of the best of all annual grasses.

*Pennisetum longistylum* — A tall, 2ft (60cm), gracefully arching plant with sparse tassels of feathery flowers which is substantial and long-lasting enough for the garden as well as good for dried arrangements in the house. Again this is less than fully hardy and is often treated as a half-hardy annual and planted out from trays.

*Polypogon monspeliensis* (Beardgrass) — A British plant that grows wild in damp pastures, often near the sea in the south, with small, silky, rather yellowish heads. Perfectly hardy and easy to raise though obviously not for dryish spots.

*Setaria glauca* (Foxtail) — A rather coarse plant with broad rough leaves and long, uneven curving heads looking rather like millet which dry to reddish brown. They are long, about 4in (10cm), and heavy and better indoors in dried arrangements than in the garden. Not the hardiest of plants, half-hardy annual treatment is preferred.

*Tricholaena rosea* — A hardy annual with spikelets covered in silky hairs in a rather ruby shade which reddens slightly as they age. It makes about 2ft (60cm) and is best dried.

**Gypsophila**    I can never understand why gypsophila seems to be so popular. When catalogues are reduced to saying that a plant is useful for mixing with other cut flowers — as they do say about gypsophila — something is clearly amiss. The perennial type used to produce gypsophila commercially is not quite so bad but the small white flowers of *Gypsophila elegans* 'Covent Garden Strain' although plentiful soon seem to curl up after cutting and they make next to no impact in the border. 'Giant White' does make some progress with flowers noticeably larger but still hardly impressive. Most of course are white but there are various red and murky pink strains which add variety if little else. I suppose the answer for the cut flower grower is to grow a lot and constantly replace it in arrangements. You will think me very unfair but this is not a plant I can enthuse over and having grown it twice, will grow it no more. If you are of the opposite opinion try 'Giant White' and also 'Colour Blend' in rose, crimson and white, 'Shell Pink' in pale pink only and 'Red Cloud' in darker carmine shades. Gypsophila is quite hardy and can be sown outside in rows in succession for cutting, but keep it out of the border.

# H

**Helianthus (Sunflower)**

Favourites for exciting the enthusiasm of youngsters in gardening simply because of their rapid growth and huge size, sunflowers are nevertheless useful garden plants although when grown to record-breaking proportions they are less than elegant. In small gardens sunflowers can look positively out of place but in broad drifts in large gardens they are very dramatic. It is a question of finding a setting of the right scale to match the scale of the plants. A row of 8ft (2.4cm) plants at the back of a 3ft (90cm) border in a pocket handkerchief garden may seem quite an achievement to a six year old, but to Mum and Dad they will more probably look like so many looming giants. The effect becomes even more strange when the first flower goes over and much smaller ones appear in the leaf joints further down the stems. So if you want to grow some to make the kids jump with glee, try and find a spot where they can be grown away from the harmony that you try to create in the rest of the garden. And pick the variety 'Russian Giant' which is specially selected for size. The ones that come by the pound from the pet shop will only produce good six-footers.

Once there were many varieties of sunflower to be had including one with variegated leaves, yet another with magenta flowers and another with enormous spherical heads. Now there are good mixtures with various yellow, bronze and mahogany shades, and a good dwarf one at about 4ft (1.2m) with vast quantities of semi-double gold flowers with black centres ('Piccolo'). The most recent addition to the range is a fully double, almost carnation-like variety which grows to only 2ft (60cm) and carries flowers 6in (15cm) across; it is called 'Teddy Bear'. All varieties prefer sun but in partial shade where the soil may retain a little more moisture the foliage is likely to look better for longer. They are all quite hardy and it is only if you want a real monster that you must sow inside. Staking is vital for taller plants and with the enormous heads becoming very heavy when wet, support high up on the plant is necessary. If you leave the seed heads in the garden, you will often find goldfinches feeding on them and they are a real delight in winter as they dart about in small flocks, their red and yellow wing flashes glinting in the winter sunshine.

**Helichrysum (Straw Flower)**

There are two distinct groups of helichrysums — those grown for their foliage, which are mostly tender perennials and those grown for their flowers.

Out of the 500 or so species which grow all over the world (except the Americas) just one, *H. bracteatum*, provides the flowering varieties that are such popular everlastings. The word 'helichrysum' comes from two Greek words meaning sun and gold which seems especially appropriate for flowers which produce such shimmering sunny shades. They are hardy annuals from Australia which until recently were grown mainly for their flowers to be used in dried arrangements. They are perfectly suited to this use and if cut and dried at the correct stage they will last for some years. The colour range is especially wide and they are easy to cultivate too.

Although they are hardy, helichrysums seem to need a fairly long growing season if they are to produce the maximum number of heads for drying and so they are best treated as half-hardy and left uncovered for the best germination. Grow them in a row in the vegetable garden or in some other area where the lack of a colourful display is not important. They will probably need staking too. The time to cut the flowers for drying is before the central yellow eye is visible, when they are about half open.

Although when dried they will last for many years, it seems to me that as soon as fresh flowers from the garden can replace them (notwithstanding the fact that good gardeners should be able to produce at least a posy of fresh flowers all the year round), they should be slung out. The taller varieties reaching 3–4ft (0.9–1.2m) are the ones usually grown for drying and they make fairly narrow upright plants which branch again from low down after the first stems have been cut. They tend to come attached to rather unimaginative names such as 'Monstrosum Double Mixed' although colours such as gold, rose, salmon, terracotta and white are also available separately. There is also a vigorous large-flowered strain sometimes described as 'Swiss Giants' with flowers as much as 3½in (9cm) across and, fortunately, with strong, stiff stems that support the flowers well; this strain too is available in a mixture or in separate colours. These varieties also make very respectable border plants. The fact that the flowers still look good after they are officially 'over' gives them a very long-lasting appeal and their rusty and coppery shades are especially attractive with red foliage and scarlet and mahogany flowers (salvias and marigolds).

In recent years the 'Bright Bikini' strain has arrived on the scene. This grows to just 15in (38cm) and makes very bushy plants. The flowers are of good size and they bring helichrysums into the realms of summer bedding for the first time. The mixture contains eight colours but it is the one colour which is usually available separately that really stands out — 'Hot Bikini'. This is a Fleuroselect Bronze Medal winner and an absolutely splendid plant. You might, though you might not, guess that the colour is rich, coppery scarlet and it looks well in contrast to white flowers like alyssum or *Dimorphotheca* 'Glistening White' as well as blending with other hot shades like *Tagetes* 'Paprika', 'Queen Sophia' marigolds and the dark foliage of *Perilla laciniata* 'Atropurpurea'. These dwarf plants can also be used to produce flowers for drying as although the stems are short, they will continue to throw more shoots from the base after the first have been cut and so, over a season, be very productive. All these varieties like plenty of sunshine and a soil which is not too rich although in good heart — good drainage is paramount.

**Heliotropum**  An old favourite of the parks department, although their propagators often used varieties raised from cuttings. 'Marine' is really the only seed-raised variety around. The plants will not be entirely uniform in height reaching anything from 12–18in (30–45cm) but the leaves are heavily veined and have a slightly bluish sheen. The large, slightly domed heads of flowers are intended to be royal purple but sometimes vary a little and some plants may be rather pale; the buds are dark. Work is going on to improve the strain and already varieties with much more lustrous foliage are being developed. There is also the added bonus of a powerful evening scent. Treated as a half-hardy annual and planted in the sun in a fertile, but not too wet a soil it associates well with orange, mahogany or, for a more dramatic show, yellow marigolds. It makes a good contrast with silver foliage and an interesting display, though not one which is to everyone's taste, with other dark and purplish flowers like *Salvia* 'Laser Purple', 'Resisto Blue' petunias and the like plus, possibly, 'Red Seven Star' marigold, *Calendula* 'Fiesta Orange', the apricot *Viola* 'Chantreyland' or you could trail canary creeper through it. If you find your plants vary too much in colour, take some cuttings of the darkest in the late summer and overwinter them for the next season in a frost-free greenhouse or frame.

**Helipterum**  An itinerant group of plants which at various times have found a home under the name of acrolinum, rhodanthe and roccardia, before finally setting up home under helipterum, although some seed catalogues still list them at what you might say is one of their old addresses. They are related to the helichrysums but are generally more dainty in habit and pastel in colour. *H. roseum* has the larger flowers, reaching up to 3in (7.5cm) across on plants 12–15in (30–38cm) high. The leaves are grey green in colour and distinctly pointed. Only double-flowered forms are available and these are the best for drying and come in a mixture and a small number of separate colours. *H. manglesii* is smaller, less grey, more delicate in habit and comes in even softer shades. The flowers are less fully double too.

As well as being good everlastings both these types are good for the front or the middle of the border. They can be sown where they are to flower and, although they close up at night and are not at their most showy on dull days, they are worth sowing in small patches where their soft colours and long flowering period make them very welcome.

**Hesperis (Sweet Rocket)**

One of the archetypal cottage garden plants but one which is not frequently seen in gardens in spite of the 'cottage garden' revival. But when we so often see that modern varieties of plants popular in cottage gardens years ago are recommended, which give an effect far more brilliant than in the days when cottage gardens were created by cottagers, then unimproved plants like sweet rocket are inevitably ignored. It is, however, a splendid plant and although strictly a perennial, is usually grown as a biennial as the first year flowers are best and it does tend to die out early in its life.

Sweet rocket is not unlike honesty in its appearance although the flowers are paler, and may be white, but the great delight of the plant is its scent which is especially powerful in the evening and at night when it serves to attract moths which pollinate the flowers. Treat it in the usual biennial manner and it will flower from June until the autumn and although it will not hit you between the eyes, it is nevertheless essential in any mixed border. It will seed itself and it only needs the seedlings to be moved to suitable spots if they have not sited themselves conveniently.

The popularity of the plant years ago is indicated by the variety of common names by which it is known. As well as sweet rocket, it has also been called dame's violet, damask flower and queen's gillyflower. It does best in a limey soil which is not too dry and it will tolerate dappled shade although it is best in the sun.

**Hibiscus**

If you want to grow some really enormous blooms then the modern hibiscus varieties are the plants to try. Some in the wild produce flowers as much as 9in (23cm) across, maybe more, on plants not more than 2ft (60cm) high — and of course with flowers of that size you do not need very many. The only species that achieves such proportions is *H. moscheutos* which grows wild in the eastern United States but which also appears sometimes in southern France; it is a half-hardy annual, and a delicate one at that. It is not a plant for standard bedding treatment, needing sun and shelter, making it an ideal container plant for the patio. Of course, in the conservatory or greenhouse it will make a splendid plant. It will grow to 1½–4ft (45cm–1.2m), depending on variety, with large slightly toothed foliage. The flowers come in a variety of red and rosy shades plus white with dark pink centres. 'Southern Belle' is the tallest at about 4ft (1.2m) but also rather slow to grow. It is hardy in a mild winter as long as drainage is good and the bed is sheltered. In its second year you will get the most enormous flowers in various pink shades or white. The best way to grow it is to prick the seedlings out into 3in (7.5cm) pots, and then into 6in (15cm) pots which can be stood outside in the summer and kept in the greenhouse over the winter. Come the spring, pot them on again, move them onto the patio in June and in July the flowering will start. Throw the party in August.

On a rather small scale are the 'Disco' varieties. There are two colours, white and rosy red and they reach about 20in (51cm). It is worth sowing them in January to try and get them to flower the same year otherwise give them the same treatment as described for 'Southern Belle'.

There are three other types of hibiscus. *H. manihot* is an Asian species, which is rather closer to a true half-hardy annual. It is tall, maybe reaching 5ft (1.5m), with pale, soft yellow flowers each with a reddish purple throat. It needs a long growing season if it is to flower outside in its first year so sow with the geraniums in January. 'Cream Cup' is a variety to look out for. *H. trionum* is the toughest of all and can be treated as a hardy annual if not sown too early. It reaches up to 18in (45cm) and produces masses of flowers, rather smaller than other types at about 2in (5cm), and yellow with a dark maroon centre. The individual flowers do not last all that long but they appear in such profusion that it hardly matters. 'Sunnyday' is the variety to look out for in catalogues. One thing to remember, especially about the large flowered types, is that they need a high temperature for germination, about 75°F (25°C), and it helps to soak the seed in hot water for an hour before sowing.

The final type is quite different as it is grown for its foliage. 'Coppertone' is one of the few really good dark reddish purple-foliaged plants with leaves rather like those of a sycamore on plants that reach about 2ft (60cm), maybe more, in their first year outside. It is neat but fairly tough and looks excellent with *Sanvitalia procumbens* or almost any marigolds up to about 9in (23cm) tall. A recent arrival on the scene and a good one.

**Humulus (Hop)**

A wonderful plant for covering new fences and trellises quickly making a good background for flowering plants. It is the variegated form with white splashes on the pale leaves which is usually found in catalogues and this is best treated as a half-hardy annual — although it is in fact a perenniial; in my garden it rarely seems to survive the winter and as it is so often grown as a temporary gap filler perhaps this is just as well. It makes 10ft (3m) with no trouble at all and can reach almost double that. As well as trellises and fences, use it to cover old tree stumps, sheds and those piles of rubble and debris which seem always to inhabit out-of-the-way corners in larger gardens. There was once a pale bronze-leaved version available, 'Lutescens', not to be confused with the yellow-leaved version of the native British hop which is a tough perennial, but this seems to have vanished from catalogues.

**Hunnemanianna (Mexican Tulip Poppy)**

Another yellow-flowered member of the poppy family, *H. fumariifolia* is a perennial, but not a reliably hardy one so is usually grown as a half-hardy annual; although as it dislikes too much disturbance it is best pricked out into pots. The plant reaches 2–3ft (60-90cm) in height and has very fine greyish foliage, which you might guess from the specific name. The flowers are brilliant, shimmering yellow and up to 3in (7.5cm) across. Bright sunshine and a well-drained soil are best and the flowers are good for cutting as well as associating with other fiery colours. It has been suggested that on especially well-drained soil it can be treated as a biennial. I have not tried this but in drier areas it should ensure an early start to flowering the next year. The variety 'Sunlite' which is at the shorter end of the height range is sometimes listed.

# I

**Iberis (Candytuft)**

Another plant too often dismissed as suitable only for children but at least this indicates how easy it is to grow. There are two species from which the varieties derive. One,

*I. amara,* is a British plant and a hardy annual which is found wild on limey soil where the competition from other vegetation is not too fierce. It also occurs occasionally as a cornfield weed, though not often in these times of the efficient use of weedkillers. This is the rocket- or hyacinth-flowered candytuft so named because the head lengthens

*Figure 7.6* Iberis *'Giant Hyacinth Flowered'*

out into a tall spike. 'Giant Hyacinth Flowered Mixed' is derived from this species and as well as being a good border plant is a fine cut flower and contains a mixture of white, red and pink. All three colours are available separately and 'Red Flash' is especially impressive in a shade that has just a touch of blue in the red and the white is a lovely plant too making relatively tall spikes on plants just 12in (30cm) high. The wild form is sometimes listed and the flowers will be a mixture of pink and white. It was

160

once used in the treatment of gout and sciatica. The flowers have a slight scent.

The one that turns up in children's mixtures is *I. umbellata* in which the flower head stays flat, even in fruit. This comes from southern Europe but occasionally escapes from gardens on to verges. It is often rather shorter, maybe reaching 6–9in (15–13cm) and it comes in a range of pale shades based on pink and white. 'Fairy Mixture' is the name usually seen and this strain is worth sowing in odd corners, cracks in paving, in the corners of gravel drives, and in spots where a little quick colour is needed. And quick colour is what you will get for candytuft flowers just a few weeks after sowing. This is another variety where the stocks vary — some are short and bushy, others are taller and more upright. You will not know what you have until your plants flower.

Sow at any time from March to the end of June and again in late August and September for some early colour in the following season. The *I. amara* group is especially useful in this respect. Candytuft can also be sown in spring or autumn to flower in pots in the cold greenhouse, pricking three plants out into a 5in (7.5cm) pot.

## Impatiens (Busy Lizzie)

For many years the busy lizzie was looked upon solely as a flowering house plant and occasionally a huge specimen could be seen totally filling a Victorian sash window. In recent times, though, with a vast effort of breeding work all over the world, many varieties which rank amongst the best of all bedding plants have been introduced. The bedding varieties have been developed from *I. walleriana*, a perennial plant from Zanzibar which, in its wild form, grows to about 2ft (60cm) with scarlet flowers 1½in (4cm) across — but it is a very leafy plant. Now the colour range has been increased dramatically, the habit much improved and the floriferousness is stunning. There is a range of plant sizes and there are even double-flowered forms and bicolours. They are not the easiest plants to germinate but they are the best of all bedding plants for shady places and for gardeners without the facilities to germinate them easily, seedlings are available from the seed companies.

The best routine for high germination is as follows. Sow in March in a peat-based compost, preferably one with sand, vermiculite or perlite added to improve drainage. It is important that trays, pots and benches be scrupulously clean, for emerging seedlings are extremely susceptible to damping off. Water the compost well before sowing and then leave it to drain for an hour or two. Sow the seed thinly and then cover very lightly. Use enough moist compost to cover the seed but not enough to exclude light. Do not water again, but cover the trays in clear polythene and then in a sheet of thin paper to exclude direct light. Milky polythene is a good alternative. The seeds need light and the highest possible humidity and of course in these conditions damping off is a danger, so hygiene is vital.

The germination temperature is crucial and should be between 70 and 75F (21–24°C). If the temperature falls below 70°F (21°C) germination will be uneven and some of the seeds may not appear. In the correct temperature germination should take place in 10–14 days. Remove the cover and then the temperature can be lowered a little, to 65°F (18°C), after germination and the seedlings pricked out as soon as they are large enough to handle. This must be done very carefully for if the roots are damaged or the point where the roots join the stem is damaged, establishment will be poor and although flowering may be advanced, growth will be very slow. The temperature should be maintained until it is clear that the seedlings are settled and starting to grow, and then be lowered to 45°F (7°C), after which the compost should be kept on the dry side to prevent botrytis infection. It is also advisable to water the pots or trays with a fungicide as a further precaution. Of course, these conditions are far easier to create in a large commercial nursery than they are in the average domestic greenhouse and even less easy on a windowsill. In practice it means starting them off in a heated propagator and if this is set to give a seed level temperature of 70–75°F (21–24°C) then many other bedding plants will be perfectly happy there too. The difficulty is usually in moving them into a high temperature after pricking out but the sort of partitioned area described in Chapter 1 will help.

Of course impatiens are actually good perennials and will root very easily from cuttings. So in theory a plant or two can be overwintered in the house or frost-free greenhouse with the geraniums and fuchsias and a stock of young plants built up from cuttings in spring. Strangely, though, it seems that not all varieties take on their normal habit when grown from cuttings and much less elegantly shaped plants are sometimes produced.

Impatiens have a wide range of uses in the garden. Initially they were especially recommended as one of the few bedding plants that would reliably produce a good display in shade — and this is still true, they remain the first choice for this situation. Their flowering capacity and tolerance of hot weather has also improved and although in the hot summer of 1984 they were not at their best, they still managed an acceptable display. They were very good in 1985. In containers they are excellent and are amongst the best of all plants for hanging baskets, both in mixtures and in single colours; on their own and with other plants. The one proviso of course is that they be kept adequately watered. The double-flowered types were initially recommended solely for use in containers in sheltered spots, in the greenhouse and in the home but I have found they also do very well outside and this is now becoming widely apparent.

One of their particular advantages as far as the home gardener is concerned is that they start to flower when the plants are still quite small so that right from the moment they are planted out they make a contribution to the garden display. Of course commercial growers can force all sorts of plants to bloom early, paying rather less attention to their later garden performance. The home gardener too can have impatiens in flower when they are planted out at the end of May, after the last frost.

One of the great problems with impatiens is, not to put too fine a point on it, that they all look the same. Now this is a slight exaggeration but amongst the varieties widely available there are just three types — large-flowered 'Blitz' types, multiflora types and doubles. 'Blitz' types, the descendants of the old 'Grand Prix', have the largest flowers, around 2in (5cm), and the plants are also a little larger than most others reaching about 12–14in (30–36cm) and making rounded plants with dark green foliage. At present the range is restricted to 'Blitz Orange' and 'Blitz Violet'. In fact the colours are closer to scarlet and deep violet. Being around the same price as the best of the multiflora type yet covering twice as much ground, you need fewer plants and so they represent excellent value for the home gardener. In fact the habit of the violet is slightly more flat and irregular than the orange. A recent introduction, 'King Kong', comes in five colours plus white and looks worth a try.

A huge majority of the varieties available now fall into

the category of multiflora types — small, low growing, spreading plants reaching no more than 6–8in (15–20cm) but carrying enormous quantities of flowers, usually a little smaller than those of the 'Blitz' type. The short internodes on most of the varieties in this range ensure a dense display of flowers. Almost every plant breeding company is developing its own strain and if you see them all side by side even the technical staff of seed companies have a job to tell the difference. So in 1983 I did a small home trial.

I grew seven of the best-selling varieties of the time, the main aim being to find out which ones were most successful in the shade. They were grown in small growing bags and set out on a shelf along a fence outside the back door. This area gets sun for about only an hour a day. The plants were started in Jiffy strips, and planted out when still quite small at the beginning of June. They all had flowers showing when planted out and they continued flowering right to the end of October — the area is sheltered from frost as well as sun.

These are the results. The heights were measured in the middle of September.

'Imp Mixed' — 18in (45cm). Too tall and sparse and the flowering was poor.

'Novette Mixed' — 14in (35cm). A splendid display of flowers in intense colours. One of the best.

'Zig Zag Mixed' — 14in (35cm). Too much red in the mixture and only an average display of flowers.

'Blitz Orange' — 14in (35cm). Large showy flowers but not all that many of them.

'Super Elfin Mixed' — 10in (25cm). Very compact and nicely spreading with superb display right through the season.

'Florette Mixed' — 8in (20cm). Turns out to be the same as 'Novette', but these were planted out from 3in (7.5cm) pots. Established less well, grew less, but the flowers were as good.

'Safari Mixed' — 18in (45cm). Tall, straggly and with a poor display of flowers.

'Futura Mixed' — 12in (30cm). Slightly trailing habit with a reasonable but unexceptional display.

You will gather that the 'Super Elfin Mixed' were noticeably better than the others as far as habit and flowering power were concerned. It was interesting to see

164

that the heights and flowering ability when grown in this very shady site were not repeated when the same varieties were grown in the open garden. Unfortunately it was not possible to grow them all in the same site but 'Blitz' and 'Imp' stood out for garden display. All grew fairly tall and in fact all but 'Blitz' and the 'Florette' (which were planted from larger pots and were slow to get going) grew much taller than their catalogue description indicated, so take all heights in catalogues and here as applying to plants grown in the open. Since this trial, quite a number of new varieties have appeared and so these recommendations include the best of both old and new.

'Accent' — Large flowers, almost the size of 'Blitz' on plants not getting much more than 6in (15cm) high but spreading to 8–9in (20–23cm). Nine pure colours with bi-colours now available.

'Novette' — (also known as 'Florette'). Contains eight pure colours and five bicolours although quite what you get in a mixture will depend on the ideas of the seed company from whom you buy it; they each have their own ideas as to the best colour balance. Generally regarded as the best variety over the last few years, support is now building up for 'Accent' and it may be overtaken. This is my standard variety for bedding and containers and it is thoroughly reliable.

'Super Elfin' — The best in shade and with the widest range of pure colours (11) but no bicolours available at all. Two colours in this range have a contrasting eye colour. Although there are a number of different pinks in this mixture, they are all quite distinct.

Other names you might come across include 'Sequins', 'Tilt', 'Amazon', 'Symphony' and the new 'Mini' series, bred for shade, which is noticeably smaller than other strains both in height and spread but will need a warm, but not too hot a summer to thrive. The red definitely makes a smaller plant than the other colours.

Bicoloured impatiens are always popular although the white star-like markings on the petals can be rather variable. 'Zig Zag' is a mixture of starred types from a variety of sources and is the most commonly seen though 'Cinderella' and 'Sparkles' also sometimes appear in catalogues. They are all at least as good in the shade where the extra brightness is valuable.

165

There are at present three strains of double-flowered impatiens available — 'Rosette', 'Duet' and 'Confection', the 'Moreton Doubles' having gone the way they deserved. 'Rosette' and 'Duet' were the first; as you might suppose

*Figure 7.7* Impatiens *'Rosette'*

'Rosette' has self-coloured flowers, in orange, pink, carmine, scarlet and shell pink while 'Duet' has scarlet and white, orange and white or carmine and white flowers. Quite how many fully double flowers you will get is open to question. It is my experience that seed companies consistently over-estimate the number of fully double flowers in the mixture. They as much as the rest of us would like these strains to be fully double but sadly this is not the case. There have been stories about the raisers of these varieties selling seed at different prices, according to the number of fully doubles it will produce, and if the plants are grown in ideal conditions they will produce more than if grown in less suitable places.

To some extent the number of fully double flowers produced depends on the way in which they are cultivated. This can be summed up by saying that the happier the plants are the more double flowers they will carry. So, avoid draughts, very wet or very dry soil, give them some shade and make sure they are fed adequately but not over generously.

Of course it comes down to the simple fact that the only advantage of these strains is that they are double. Although 'Confection' has the best habit, generally the habit is less good than that of the multiflora types and fewer flowers are produced too. So unless you are prepared for three-quarters of your plants to carry single or semi-double flowers, forget it until some better varieties appear. At present 'Confection' is probably the best available — although it produces no more fully double, it does carry more semi-double and fewer singles than other varieties.

Only two open-pollinated varieties are worth growing. There are a number of others about but none compare with the F1 multifloras. These two are the only New Guinea types available from seed. 'Tangeglow' is an excellent variety, once recommended solely as a pot plant but proving an excellent bedder and container plant too. As you will have guessed the flowers are a very unusual tangerine colour and they are even bigger than those of 'Blitz' at 3½in (9cm). The plants only grow to about 6–8in (15–21cm) but spread to twice that making open plants with the flowers carried individually on long stems. Very unusual and worth trying. 'Sweet Sue' is newer, it has dark foliage with a reddish tint and with reddish stems too. The flowers are not quite as big as those of 'Tangeglow' but the colour is slightly more intense. 'Sweet Sue' makes a much smaller plant outdoors and needs a favoured site and season to thrive but in the cold greenhouse makes a much more manageable plant than 'Tangeglow' which grows rather large. 'Tangeglow' was Highly Commended in RHS trials in 1975.

Most of the 'New Guinea Hybrids' are at present available only as young plants raised from cuttings although more seed-raised strains are at present under trial. Their unique quality is that they combine large flower size, a wide range of flower colour and prolific flower production with variegated and coloured leaves. The leaves are longer than those of the familiar types and the stems altogether more stout and fleshy. Depending on variety they make large, rounded plants up to 18in (45cm) high or spreading plants 9in (23cm) high and nearly 1ft (30cm) across. The leaves may be rich purple, fresh green with a long yellow splash, green with a yellow splash and scarlet veins or any of a number of other interesting combinations. The undersides often contrast dramatically with the upper surface. Flowers come in scarlet, various pinks and lilac and the overall display is stunning.

Again, a strange timidity was demonstrated when these first appeared and they were recommended for cold greenhouse and indoors only. In fact they thrive in containers and also make an excellent show outside as long as they are sheltered from cold winds which will soon finish them off. Indoors be especially vigilant in searching for red spider mite and dealing with it promptly for even more than the familiar types they will go down rapidly once infected — as I know to my cost.

In the garden, mixtures of busy lizzies are most definitely best used in isolation from other plants. They are ideal in small beds of their own or in spaces in mixed borders contained and bordered by other plants. They can be bordered by other bedding plants, but in single colours, or by silver, green or yellow foliage please. And do not set them in front of a dwarf antirrhinum mixture and alongside a mixture of 'Quikstep' mimulus. This is an ideal combination if you feel that life is altogether too dull and straightforward for it will scramble your brain as soon as you look at it and suddenly ordinary aspects of life will seem quite bizarre. Even marigolds will seem interesting. Impatiens are ideal planting for shady northern backyards and the sort of nebulous, rather useless areas outside the backdoors of Victorian terraced houses. They thrive in containers and bring a splendid brightness to dark and dull corners. Your entire summer can be transformed by a few troughs or growing bags and some hanging baskets filled with 'Super Elfin' or 'Novette' impatiens. The bicolours are especially good here although most modern mixtures contain sufficient whites, pale pinks and lilacs to counterbalance the darker reds.

**Ionopsidium (Violet Cress)** The smallest of the annuals described in this book, reaching just 2in (5cm), with tiny lilac and white flowers in large numbers. This is an annual that can be left to seed itself on the rock garden and in raised beds and will rarely be troublesome while producing flowers whenever the temperature is not below freezing — and sometimes then too. It also makes a lovely plant for the cold greenhouse in winter, when seeds sown thinly in a wide pan in the autumn will grow and flower right through, eventually reaching all of 6in (15cm) high and flopping over the sides. The leaves tend to go rather yellow by this stage though. Sow it, too, in cracks in paving.

At Kew it has even been used in spring bedding schemes, without a great deal of success, but in small beds with clumps of *Iris reticulata* or some white *Crocus chrysanthus* such as 'Snow Bunting' it would be very pretty; the scale must be small though.

The climbing morning glories include possibly the most beautiful and the most hideous plants described in this book. 'Heavenly Blue' and 'Roman Candy' they are called — eulogies and vitriol follow soon. The ipomoeas now seem to include Quamoclit, the climbing annual Convolvulus and Calonyction. They cover a range of half-hardy climbers for sunny sheltered sites in soil which is inclined to retain some moisture all summer — an almost impossible quality without recourse to the sprinkler. All have flowers like the greater bindweed. The colours of the cultivated forms vary greatly from pure white, through sky blue to dark blue and pink, lilac, purple and scarlet. Most are actually perennials but are not hardy outside although in a greenhouse or conservatory they will overwinter happily — which produces another problem in that they grow so furiously during the season that there is hardly room for anything else. So half-hardy annual treatment is the norm, although it does have rather special requirements.

**Ipomoea (Morning Glory)**

Soak the seed overnight before sowing. Put the seeds in a saucer with a little water and the label and leave them in a room which does not get too cold at night — a shelf near a radiator or boiler is a good spot. Next day, sow the seeds individually in 3in (7.5cm) pots of peat-based compost with extra perlite or vermiculite. They need a high temperature to germinate — 75°F (24°C) is about right but after germination this can be reduced. It is better not to keep them warm as not only will they grow so much that you will be forever pinching them out, but they will also need even more careful hardening off than usual.

Hardening off is very important. Plants which are not suitably hardened suffer badly when planted out, especially if they go in a draughty spot. The new leaves get smaller and all new growth takes on a particularly anaemic appearance, almost white. Often they take some weeks to recover and in severe cases do not recover at all. So grow them cool, harden off steadily and thoroughly and choose the site well; draughts, and worse, winds, are not helpful — sun and shelter are. This is a case where a good soak with a liquid feed straight after planting is very useful. They all need support and tall hazel twigs, plastic mesh, trellis or best of all a suitable shrub are ideal. They are rarely successful grown on the tubes of wire that are so successful with sweet peas.

Now, the varieties. This is the bit you have been waiting for. 'Heavenly Blue' is exquisite, quite perfect. Summer sky blue in colour, softening to white in the centre of the large, open flowers, everyone who sees this plant is captivated. Every home in the country should be sent a free packet each spring on the National Health — it would do wonders for the morale of the nation in these hard times. At its best growing through other plants, try it twining through an overwintered *Eucalyptus globulus* where the silver foliage makes a perfect background and it also looks surprisingly good climbing through a mature *Eccremocarpus scaber* — the standard orange strain. Last year I tried it through the rose 'Mary Rose' — a strong pink variety and one of the perpetual flowering 'new old roses' from David Austin. This year I intend to try it up an *Onopordon arabicum* in its second year — should be lovely.

'Roman Candy', also known as 'Minibar Rose', is one of the ugliest plants ever bred, unrivalled in its expression of an excess of bad taste. It is a small-leaved thing, less vigorous than 'Heavenly Blue', thank goodness, and of a singularly sickly appearance owing to the leaves being so splashed with white that less than half the leaf surface is actually green. As if this were not bad enough there are the flowers. These rival the geranium 'Picasso' in their ugliness being a singular shade of cerise-magenta with a white eye and a white picotee. We can only be thankful that they are not bigger — although then I suppose they might hide the foliage, which would be no bad thing. But then if the leaves were larger they might hide the flowers. Talk about two evils. If I see this in your garden I shall be over the fence in a flash and it will soon be ripped from the soil. Be warned.

Other varieties excite less controversy, and all are worth growing; 'Flying Saucer' has flowers in pale blue with white streaks and 'Sapphire Cross' is similar. 'Scarlet O'Hara' has brilliant red flowers which are bigger than most and very dramatic. 'Early Call' also has big flowers, up to 4in (10cm) across, in a range of colours including reds, blues, pinks and some very dark shades. The plants come into flower earlier than most varieties and so this is an especially good variety in cooler parts of the country.

# K

Dumpy, bearskin-like bushes in fresh green which turn to rusty or purple shades in the autumn. Like annual versions of dwarf conifers, these plants are probably best grown as an annual hedge rather than dotted through annual schemes where they can often look rather odd. They are easy enough to raise in the usual half-hardy manner although they germinate best if the seed is not covered. They grow from 2–4ft (0.6–1.2m) high, depending on variety, and for a hedge effect are best planted out around 9in (23cm) apart. If you want to grow them as single specimens leave a couple of feet between the plants to get the benefit of their unique shape. They look remarkably spindly and thin when you plant them out but quickly bush out. In high winds they can topple so do not put them in exposed spots. They are happy in most garden situations as long as the site is not too shady or the soil too wet.

**Kochia (Burning Bush)**

*K. trichophylla* is the variety usually grown and this reaches 2ft (60cm) with fresh green ferny foliage which turns scarlet and bronze in the autumn. 'Acapulco Silver' is a more recent introduction which is rather taller at 3½ft (1.06m) and has its foliage speckled with white. This variety turns a very attractive deep purple. Both will keep their autumn tint well if dried and so are very useful in winter arrangements.

The shape of the burning bush plants usually causes a smile as they are so unusual and to accentuate this feature make sure they are surrounded by something shorter and fairly flat growing. If you prefer to play the dumpiness down, group three together with a more tall and uneven underplanting.

# L

There has been an enormous number of books written on sweet peas over the years (two of the best are listed under Recommended Books in Appendix V) and, as most of these are concerned with growing sweet peas for exhibition, this is a field which I do not intend to cover. However, that still leaves plenty to write about.

**Lathyrus (Sweet Pea)**

The wild original of the sweet pea comes from Italy and can still be seen in botanic gardens. It has a very small, but intensely coloured flower and a powerful perfume. It is still worth growing the first descendants of the wild sweet pea as the scent is quite special and a few of them amongst some of the less scented varieties really lifts them.

The simplest way to grow sweet peas is up trellis, walls or a fence. A south-facing wall in full sun is the worst spot to grow them as the soil is dry and the heat from the wall encourages the blooms to go over quickly — try a west wall for preference, east or even north. There are enough climbers that prefer a south wall so do not despair.

The soil must be well prepared. Sweet peas are classic examples of a maxim that applies to many annuals. They will produce a display of sorts in almost any conditions but for the best display they must have their needs catered for. So take out a trench the width and depth of the spade, fork in some organic matter, then refill the hole mixing some more muck into the garden soil too. If you want to be extra sure, a dusting of bonemeal forked into all levels will help. Make sure it is all well firmed but leave the final level a little below the surrounding soil and with a ridge about 2in (5cm) high along each edge. Seed can be sown in mid-March where they are to flower. Not everyone agrees, but I still think it's worth soaking the seed before sowing; it should advance flowering by about a week. First, with a nail file, rub the coat until the creamy green seed is just visible and then put the seeds in a saucer with a little water overnight. By morning the seeds will have swollen and they can go outside. Make sure the soil is moist, and set the seeds in dibber holes 1in (2.5cm) deep, 3in (7.5cm) apart. A mouse trap is a sensible precaution.

Support on fences and walls can be a problem and it pays to get it in place after roughly preparing the soil but before sowing. Plastic clematis net is probably the simplest but is not all that elegant. The brown is probably the least offensive and it lasts for years. You can get special clips to fix it to brickwork or timber. Pea sticks are probably the most aesthetically pleasing but are not now easily available. Try and get hazel if you can, and tie it into the wall otherwise the whole lot will blow down when the display is at its most perfect. Canes are not satisfactory against a wall or fence — there are just not enough of them. Strings work perfectly well but you need so many hooks or nails that you might just as well put up some trellis which is by far the most elegant but also the most expensive. A few pieces of fine twig at the bottom to help the seedlings on their way is very sensible. At about 3in (7.5cm) pinch out the tops and away they go. Keep an eye on them and if they decide to make for the path rather than the fence, guide them in the right direction. They will look after themselves and by June will be in flower.

The alternative is to sow in the autumn. You will get them another couple of weeks ahead by autumn sowing but in bad winters the survival rate can be rather low. If you really want to be super efficient then sow in the autumn in a cold frame and overwinter before planting. Five seeds around the edge of a 5in (12.5cm) pot is about right; it is important not to keep them too sheltered over the winter and, at the same time, not to let them stay frosted in their pots for weeks. A cold frame is ideal — mats on the glass in the worst weather and open lights on warm days is the right approach.

In March, knock them out of their pots and plant firmly so the stems stand upright. Do not forget the slug pellets.

Dead-heading as the plants grow and the flowers fade is the one vital task from here on — leave the dead flowers on and you will soon have a fine crop of unpredictable seeds and no flowers. So when the last petals go limp, take off the flower stems with the secateurs at the leaf axil. If you want to do your best by your plants then a liquid feed every couple of weeks from the time they start flowering would be in order. Dahlia or chrysanthemum feed is ideal or whatever you happen to have as long as it is not a high nitrogen lawn feed.

The one big problem with sweet peas is their tendency to give up the ghost some time in August. It is asking rather a lot of them to be so prolific for so long but gardeners are demanding characters, especially annual enthusiasts, and so we must try and recommend something to prevent a mass of dry brown foliage disfiguring the fence over the August bank holiday. The first thing, after the dead-heading, is regular watering and feeding; drought kills. It has been suggested that by cutting the plants back in early August and watering and feeding well, that they will dramatically revive. Have none of it; all you end up with is half a mass of dead foliage. Personally I think that a second sowing in May amongst the existing stems is the answer. It is important to keep the soil moist, but this should lead to a new flush of flowers taking over as the old ones fade.

Of course there are more ways of growing sweet peas than against fences. In modern, smaller gardens where space is at a premium, as well as in larger demesnes, sweet peas look splendid grown through shrubs. Rather than sow underneath it does make sense to put in young plants so that the seedlings do not suffer from too much root competition. It is usually impractical to prepare the soil as thoroughly as you would for plants growing against a fence without damaging the roots of the shrub that is acting as support, but do your best. You will need a few twigs to get the plants off in the right direction.

There are two approaches in choosing the plants through which to train the sweet peas. You can pick plants like forsythia and weigela which flower earlier and extend the contribution that their space makes to the garden picture by using them to support sweet peas. Alternatively, choose plants like shrub roses with which sweet peas will make attractive combinations. *Continus coggygria* 'Royal Purple' is a splendid host for a long-stemmed white variety such as 'Diamond Wedding' and the new pink English rose from David Austin, 'Mary Rose', looks delightful with the blue 'Noel Sutton'. This is a strong combination that not everyone likes so find your own combinations by picking a few flowers of your favourite varieties and trying them side by side with the shrubs you happen to have in flower.

Another way of growing sweet peas in borders is to construct free standing supports which can be placed at strategic spots amongst permanent or entirely annual

*Figure 7.8 Plant sweet peas on free-standing netting supports to give height and interest to borders*

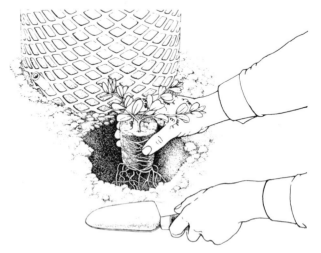

plantings. The wigwam of half-a-dozen canes, such as is used for runner beans, works well but suffers from the singular disadvantage of getting narrower at the top — the very area where the sweet pea is likely to be at its bushiest. A splendid system seen at Glasnevin Botanic Gardens in Dublin and also in other gardens is to nail a tube of chicken wire to a stout stake which is driven into the ground to support the cylinder. Clematis netting comes in just the right width to make a suitable sized tube: 6ft (1.8m) of net pinned to the top section of an 8ft (24m) post gives enough to knock into the ground to keep it stable. Five sweet pea plants are planted together, straight from their pot, and the net carefully knocked into place around them. For a wigwam, five or six 8ft (2.4m) canes are knocked in and the tops tied giving about 9in (23cm) above the tie, one plant goes by each cane. In both cases thorough preparation is important with the usual muck and bonemeal added.

Growing sweet peas for cut flowers is a different thing altogether but worth the effort. I have grown them on the cordon system, as they are grown for showing, and naturally up netting and I am sorry to have to say that the extra time and trouble of cordon training seems to be worth it. The length of stem and the number of flowers per stem as well as the size of the flowers is noticeably better. With the natural system you certainly get a lot of flowers, but although the first few can be very good the standard later really is not up to scratch. The details of the cordon system are found in most sweet pea books.

Alternatively, try the natural method. Prepare well, as you would for runner beans, knock in the 8ft (2.4m) stakes at each end and fix a stout wire to the top of one post. Thread it through the topmost mesh of some strong nylon pea and bean netting and fix the wire to the top of the other post. Run a similar wire along the bottom for stability. Plants are set out every 6in (15cm) and left to climb. Vast numbers of flowers can be produced by this system — I counted 670 from 21 plants in 10ft (3m) of row by the end of July, but a lapse in watering and feeding while on holiday heralded the end of the display and the stems were getting very short by then anyway. 'Leamington', 'White Leamington', 'Lady Fairbairn', 'Sheila McQueen', 'Noel Sutton' and 'Southborne' did especially well in that trial, especially in stem length which is the one thing that seems to suffer most under the natural system.

These twelve varieties cover most of the colour range:
'Beaujolais' — A rich dark maroon which is better as a cut
flower than in the garden.

175

'Mrs Bernard Jones' — Large flowers in almond pink on a white background.

'Lady Fairbairn' — An unusual delicate pinkish lilac shade with long stems.

'Leamington' — An old variety with frilly lilac flowers.

'Midnight' — A recent replacement for the rather short-stemmed 'Black Prince', it has the deepest dark maroon flowers.

'Noel Sutton' — A long established favourite amongst the blues, the colour is rich, and the stems long.

'Red Ensign' — The best of the bright reds, this is an intensely scarlet shade.

'Royal Wedding' — Very long stems and pure white.

'Sheila McQueen' — Orangey salmon flowers that do not fade in strong sun.

'Southborne' — Pink on white with very stout stems; an old favourite but still good.

'White Leamington' — Large frilly flowers on strong vigorous plants.

'Wiltshire Ripple' — Chocolate stripes on a white background; a must for the adventurous flower arranger.

Not all of these varieties are strongly scented so if scent is the real priority try these: 'Royal Wedding' (white), 'Cyril Fletcher' (pink), 'Ballerina' (dark pink picotee), 'Old Times' (cream flushed with blue), 'Evensong' (bluish lilac), 'The Doctor' (mauve).

Apart from the Spencer varieties which I have been dealing with up to now there are a number of other tall types that are worth growing.

*Royals*

The Royals were developed from the old Cuthbertson floribundas in America. The Cuthbertson floribundas arose from crossing Spencer types with an early-flowered American variety. Their great advantage was that they flowered a fortnight earlier than the Spencers and so were valuable for growers wanting to get a good price for the first flowers to reach the market. However, the range of colours was poor and the individual flowers less attractive. The Royals flower between the Cuthbertsons and Spencers. The stems are longer, the individual flowers more shapely as well as larger and they are altogether more vigorous. They can be grown under glass for early flower arrangements or outside they will still come into bloom before the Spencers which make up the bulk of varieties available.

Flowers far earlier than the Spencers and the flowers are very elegant and well arranged on the stem and of good size. There are about a dozen varieties available but most seed companies sell only a mixture. This group is probably the best for flowering in the cold greenhouse or polythene tunnel.

*Early Multiflora Giganteas*

The important characteristic of the galaxys is that they carry eight flowers on a stem instead of the usual five. There is a good range of colours and they are ideal varieties for powerful displays in borders where the large number of flowers makes a substantial difference. They are not suitable for cordon-style growing or for growing naturally as cut flowers. The problem is that when the top bloom opens the bottom two or three are passed their best. Again, many companies only list mixtures but you will find that specialists list individual varieties.

*Galaxys*

Many plant breeders, particularly at Ferry-Morse in America, have developed varieties which are shorter than traditional types. There is now a range available ranging from 4ft (1.2m) down to very dwarf types at about 6in (15cm).

'Jet-Set' — Almost two dozen distinct varieties have been raised, but in Britain only a mixture is usually available. They reach about 3ft (90cm) in height and carrying about five flowers on stems longer than those on other dwarf types. 'Jet Sets' can be left to sprawl in the border but for cutting and in windy sites where more control is needed, short pea sticks are useful. Too big and unruly for containers, except really big urns where they can be left to tumble elegantly, but good in mixed borders where they will peep in and out of woody plants without overwhelming them.

*Intermediate Types*

'Knee-Hi' — Maybe a little taller than the 'Jet-Sets', especially if sown in the autumn, but that little bit less impressive.

'Snoopea' — An unfortunate name which has been the subject of legal wrangles, temporarily disrupting the more or less friendly relations between the British seed companies. Its original name, 'Persian Carpet Mixed', may be rather a cliché and uninspiring but it does not cause the cringe of 'Snoopea'.

It is a good variety reaching about 2ft (60cm) in a bright range of colours with the scarlet usually available separately in Britain. The plants are grown easily without support, they have no tendrils and the stems are usually long enough for cutting. It has an unfortunate tendency to produce rather too much of its narrow foliage unless grown in the maximum of sunshine. The arrival of a successor called 'Supersnoop' has, mercifully, received rather less publicity than greeted the arrival of its forebear but is virtually indistinguishable apart from the fact that it is said to flower ten days earlier. In spite of the name it is an excellent garden variety which, unfortunately, does no better than most shorter sweet peas when grown in baskets. The root space is too restricted, I fear. 'Snoopea' makes an excellent sole occupant of a narrow path-side border.

*Dwarf Types*      These are the ones for containers and corners of borders. Delightful little plants with the flowers slightly smaller than their taller relations and so maintaining a happy proportion.

'Bijou' — A good window box variety where it trails prettily, in the garden it usually makes about 12 or 15in at most (30–38cm). The colour range is good and the growth stout and stocky so no support is necessary. The flowers are especially frilly.

'Patio' — No more than 12in (30cm) with especially scented flowers produced in vast quantities. Although only about four flowers are carried on each stem, the display is very impressive and they are excellent in pots and window boxes and even in growing bags.

'Cupid' — Unique in its habit of growth, 'Cupid' reaches only 6in (15cm) high but spreads out well, making a flat carpet covered in short stems each carrying two or three flowers. This variety was grown a little in the 1930s but then disappeared and has only recently resurfaced in New Zealand from where it was re-introduced into Britain. Charles Unwin, whose firm still specialises in sweet peas, writing in the late 1920s reported that 'Cupid' was little grown owing to the fact that in poor summers the flowers tended to drop off before they opened. He revealed, though, that the type was discovered in 1894 by Morse and Company, now partners in Ferry-Morse who are still working on dwarf sweet peas! 'Cupid' comes in a mixture of bicolours involving various reds, pinks and lilacs — all with white. All dwarfs seem particularly happy in well-drained, fairly fertile soil in bright sun but 'Cupid' especially so. Well worth growing in the right situation.

Two important things need to be said about all these dwarf varieties. First, they tend to have especially hard seed coats, all of the colours, so it pays to spend your winter evenings in front of the snooker on the TV filing the seed coats of even those that you intend to sow direct outside. Secondly, although very large numbers of flowers will be produced over the season, dead-heading is essential otherwise the flowering will stop, the plants will die and you will have to find something else to fill an ugly gap in the border.

And a final note on all varieties. They do not like malathion so choose a different chemical if you have pest problems.

**Lavatera (Mallow)**

The lavateras include, apart from some especially tenacious weeds, one of the best new varieties of annuals introduced in recent years, 'Silver Cup' — an indispensable plant which ought to be grown in every garden. The first Fleuroselect Silver Medal winner, it is a hardy annual to sow in spring or autumn which grows to about 2ft (60cm) making bushy plants branching from low down if thinned to about 15in (38cm). The leaves are not dark but the flowers are stunning. Big, soft pink, open bells up to 2in (5cm) across with dark veins, they appear from mid-June to the autumn. Lavateras like sunshine and any soil which is reasonably fertile and well-drained. The only problem is that in hot dry summers they tend to give up flowering rather early in the season leaving a singularly unattractive clump of dead twigs. So soil that retains a little moisture helps. Ruthless thinning at the seedling stage will encourage branching low down to give a succession of flowers.

There is another excellent variety, rather shorter with much darker foliage called 'Mont Blanc' which as you may expect is pure white. Unfortunately its true pink counterpart 'Mont Rose' is very rarely seen and yet 'Silver Cup' and 'Mont Blanc' are not a great success together because of their differing sizes and leaf colour.

*Figure 7.9* Lavatera
*'Silver Cup'*

'Silver Cup' is ideal in the favourite pink, blue and silver schemes with tall or short ageratum, various silver foliage cinerarias and pyrethrums and maybe white petunias and *Salvia farinacea* 'Victoria'. 'Mont Blanc' makes a splendid contribution to the white border with 'Iceberg' or 'Margaret Merrill' roses, one of the silver cinerarias like 'Silverdust' plus *Helichrysum petiolatum*, white petunias and some fresh bright green parsley like 'Bravour' — plus maybe a plant or two of *Atriplex hortensis* 'Rubra' — which I suppose makes a sort of whitish border.

If you want to raise these varieties as half-hardies to get them out and into flower a little earlier, then sow in March but grow cool and plant out in early May if well hardened off. Both varieties can also be sown in September and overwintered in pots in a cold greenhouse. It pays always to prick out lavateras into pots as they resent root disturbance and this can curtail flowering.

There is another lavatera which is sometimes seen, a plant which grows wild on the coasts of the British Isles, even quite near to popular beaches such as at Skerries a little way north of Dublin. This is *Lavatera arborea*, a woody biennial that makes a stout fatly branched plant with big pink bells, darker veined. Not far away from Skerries at Malahide Castle I first saw the extraordinary variegated variety which is one of the few variegated plants to come true from seed. The flowers are deeper in colour and the variegation rather irregular and creamy. It seeds itself about the garden happily and once you have it you are never likely to be without it unless a very fierce winter kills off all the overwintering plants. Well-drained soil, as you might expect from a plant that grows on shingle and sand, is a great help to sturdy growth and successful overwintering.

**Layia (Tidy Tips)**

A rather frivolous common name for a pretty little hardy annual in the daisy family. Rather variable in its natural home in North-west America the form in cultivation grows

*Figure 7.10* Layia elegans

to 18in (45cm) and has 2in (5cm) flowers in bright yellow, each of the rays being tipped in pure white. The discs are dark yellow. The foliage is fine and feathery and although the plants have rather a willowy appearance they seem to support themselves fairly well unless grown on very rich soil. The flowers last well in water too. This plant looks either stunning or ghastly, depending on your taste, when grown behind the poached egg flower, *Limnanthes douglasii* which has similar flower colour.

**Leptosiphon**

Also known as gilia this tiny hardy annual reaches just 6in (15cm) and often slightly less. The foliage is very narrowly lobed and the flowers, although small, are produced in large numbers and in brilliant colours. Sun and a reasonably light soil will suit them well and they can be sown in empty spots on the rock garden or raised bed, in the front of the sunny border and in window boxes and tubs. They are also one of the few plants that will thrive sown along the cracks in paving bedded on sand. Mixtures appear under such names as 'Rainbow Hybrids' and the six-petalled flowers, each with a yellow-orange eye, come in white, cream, yellow, orange and various reds and pinks.

**Leptosyne**

Whether you look for this under the name used here or under coreopsis depends on whether you are a botanist or a gardener; strictly speaking they all belong under coreopsis. The only variety you are likely to come across is *Leptosyne stillmannii*, in particular its variety 'Golden Rosette'. This is another yellow daisy-flowered plant, growing to about 18in (45cm) and a good cut flower with dramatic, double, dark yellow daisies above dense, rather narrow foliage. Treat it as a hardy or half-hardy annual, but be careful when planting out if you treat it as a half-hardy as this is another plant whose flowering can be curtailed by rough treatment when young.

If you should come across *L. douglasii* in a catalogue, give that a try too. It is rather shorter and more lemony in shade with single flowers and is quite compelling — especially as it flowers for so long.

**Lilium (Lily)**

In some catalogues you will find varieties listed which it is claimed will flower in their first year after sowing; 'Snow Trumpet' and 'White Swan' for example. The idea is that seed is sown in winter, pricked out into individual pots, planted out in spring and the plants flower later the same summer. I have never been able to get them to do that. I sow in early January at about 70°F (21°C) and although I had excellent germination and no losses and they seemed to grow well, there were no flowers until the following year and even then they were not very large. Maybe if seed were sown in November, kept growing well right through the winter and planted out as more substantial plants it may work but by the time you have spent all that money on keeping the greenhouse warm in the depths of winter you may as well buy some fat bulbs. In warmer parts of Europe

182

and the southern United States, they might be more successful.

**Limnanthes (Poached Egg Flower)**

Also known as fried eggs, the flowers as you might expect are yellow with at least half the length of each of the five petals in white. The plants only grow to about 6in (15cm), the leaves are freshly coloured, finely divided and rather fleshy and the plants flower for many, many weeks. They seed themselves about easily and turn up in many useful spots in the garden. They prefer a cool, moist root run and so do well alongside paths, and at Cambridge University Botanic Garden they look lovely and thrive amongst rocks overhanging a stream. Once upon a time there were more variants available — a white, a white with pink veins, a large-flowered form and a pure yellow form which is grown at Kew but which does not seem to be available from seed companies. Look out for them. Ideal plants for walls, in containers too and window boxes. An accommodating and cheery little plant.

**Limonium (Statice)**

One of the widely grown everlasting flowers, it has rather less showy relatives that grow in muddy salt marshes all around the British coast. Fortunately, the cultivated statice (*Limonium sinuatum*), a Mediterranean plant originally, is rather less fussy in its requirements. It grows 15–30in (38–75cm) high and the fresh green stems have flat wings running along them and pale green divided foliage. Treat statice as a half-hardy annual sown in late March or as a hardy one to be sown in late April.

One modern variety stands out in comparison with older types. The 'Fortress' strain from Holland grows to 2ft (60cm) but is very stout and well-branched so makes as good a bedding plant as a plant for drying. It is available in six colours, including some strong and unusual apricot shades, as well as in a mixture and in the border it lasts especially well in addition to starting to flower early. Grow rows in a special plot if you want plants for drying and cut them when most of the flowers are open. Hang them upside down in a dry, airy spot; they are amongst the easiest to dry and it is sometimes suggested that the green wings be stripped off the stems before they are hung but not everyone can be bothered. They do best in full sun in a well-drained soil that is fairly fertile and can also be grown as spring pot plants if you have a frost-free greenhouse: for this purpose they should be sown in September. Cut stems as they become ready and from early sowings a large number of stems can be cut over the season.

If you find the 'Fortress' strain difficult to come by, you might come across 'Blue River'. This is even shorter, at about 18in (45cm), and is a slightly variable shade of blue although most will be quite dark. Various other strains appear, often simply called 'Mixed'.

**Linanthus (Mountain Phlox)**

Another plant in the same group as leptosiphon which is often put in with gilia. *L. grandiflorus* is a hardy plant with finely divided leaves on plants reaching 2ft (60cm) in height. The flowers are just an inch (2.5cm) across, rather funnel-shaped and are lavender purple in colour with white flecks. The usual open, sunny and fairly well-drained conditions are ideal although in partial shade it will still flower well. It has a long flowering season, is a good cut flower and altogether an elegant and airy plant.

**Linaria (Toadflax)**

Only the one annual, *L. maroccana* is commonly grown and a delightful dainty little plant it is too. Tiny snapdragon-like flowers are carried on short plants which branch well if thinned out to 6in (15cm). The flowers come in a wide variety of red, purple, pink and similar shades with yellow or white throats. Good in little patches in gaps in the border, in cracks in paving or wild corners of gravel drives, they will seed themselves into, and along borders, over the fence to next door or between the pavement flags — but they are never so plentiful as to be a nuisance. Like the 'Bowles Black' viola which can be left to sow itself at random and simply removed where it is not wanted, the same can be done with these little toadflax. If only they could be had in separate colours — a constant moan, as you will no doubt appreciate by now! 'Fairy Lights', 'Northern Lights', 'Fairy Bouquet' and other similar named varieties (but not 'Northern Bouquet') are the ones to look for.

Rather different is 'Crown Jewels' a 9in (23cm) variety of the rather tall *L. reticulata*. The colours are brighter than the various 'Fairy' forms. Maroon, orange, red and gold flowers with mainly yellow throats — very pretty. Any soil will do and as long as they get sun for part of the day they will be quite happy. Like a lot of hardy annuals, good thinning is important for long flowering as is a less than scorching summer. In 1985 my 'Northern Lights' were one of the first hardies to flower and lasted almost the longest.

**Linum (Flax)**

There are two splendid plants, quite different in colour. *L. usitatissimum* is the flax that used to be grown in large quantities for linen and linseed oil. Not long ago I came across a field in full flower as I drove through Northamptonshire and the sight was so stunning I nearly ended up in the ditch. I suspect that this crop was grown for linseed oil which is still used to treat cricket bats and outdoor timber furniture. This same variety is available to the home gardener and a delightful plant it is with soft, sky blue flowers on greyish, rather willowy plants reaching about 18in (45cm). Grow it in large clumps in the annual border or let it seed itself in mixed borders, moving the seedlings about as necessary.

The other linum grown is quite different, *L. grandiflorum rubrum*. In its native North Africa it grows to about 3ft (90cm) but the form in cultivation is more modest at about 12in (30cm). The colour is the most extraordinary blood red with a slight glistening sheen and the flowers are up to 1½in (4cm) across.

Sun and well-drained soil are necessary for both I am afraid and if you can be bothered, dead-head regularly to ensure a long-lasting crop of flowers. The red variety looks well against purple sage or any other foliage in a similar colour and is also good peeping through silver foliage. The tall blue is fine peeping through anywhere in the mixed border.

**Lobelia**

Famous for its alternation with alyssum in front of scarlet salvias there is, fortunately, a great deal more to the dainty little lobelia than that. Naturally, it is a tender perennial which is treated as a half-hardy annual and which over the years has become segregated into two groups — the bedding types and the trailing types. There is also at least one variety which does not come true from seed and which must be propagated from cuttings each year and overwintered. The ease with which lobelias root can be seen from the fact that towards the end of the season many plants in borders produce a dense covering of short white

185

roots towards the base of the shoots where they are not exposed to the sun. These root very easily and for gardeners who like to try something a little out of the ordinary, bedding types can be rooted and overwintered along with the geraniums.

The seed of all lobelia is very small and so is often difficult to sow, especially for the less experienced. The seedlings are also very small and not easy to prick out. However, seed is not expensive and so it is perfectly feasible to prick out the seedlings in small clumps. Treat it like begonia and leave it uncovered.

The bushy bedding types come in a variety of shades going from 'Snowball' which is white but flawed with 10 per cent blue-flowered plants through 'Cambridge Blue' to 'Mrs Clibran Improved', a very deep colour with a white eye and 'Emperor William' which is without the eye. There is also 'Rosamund' which is a deep slightly cherryish red. There are one or two good mixtures as well. They like a soil that does not get too dry which is why they always did so well in those parks department displays of years ago, when the beds were sometimes manured twice a year and were so full of humus that they never dried out thoroughly unless the weather was quite exceptional. So give them good soil, and sunshine if possible.

All the colours are useful in making pictures in the garden. Try 'Cambridge Blue' with a short white nicotiana such as 'Domino White' and a pink geranium such as 'Pulsar Rose', 'Appleblossom Orbit' or 'Gala Rose'. In a small bed try it with 'Video Rose' grown from cuttings of plants from the 'Video' mixture. The darker blues like 'Crystal Palace', which also has the benefit of dark bronzish foliage, make a striking show with a good yellow like *Calceolaria* 'Midas' or a slightly lemony shade such as *Pyrethrum* 'Golden Moss', the taller of the yellow foliage pyrethrums. In larger areas try the addition of the slightly paler though very intense *Salvia patens*. And what about *Cineraria* 'Silverdust' and *Salvia farinacea* 'Victoria'?

Then there are the trailing types, usually recommended for baskets and window boxes. Personally I am not sure that this group really is ideal for baskets. I find that if the baskets are well looked after, especially when it comes to watering, the lobelia gets far too long and dangling to be really effective. The same applies if the baskets are set in an especially sheltered and warm spot such as under a glass porch. So I use the bushy types instead. But in cooler areas you may find that the trailing types really are best.

There is a similar variety of colours available including perhaps the nearest colour to red that is available in all lobelias. Mind you, in spite of the seed company's publicity, it could not be described as really, well, red. 'Ruby Cascade' may be nearer carmine than the more or less purple 'Red Cascade' but think of it as red and the plans really will go awry. The other problem with lobelia in baskets is that they tend to burn out before the end of the season if not carefully looked after. And there is nothing worse than great brown gaps appearing in your displays in the middle of August. Think of it another way, though, and this tendency can be used to advantage. If you plan for them to be no more than fillers, and they are quick growing, then they will carry a heavy burden of the display until the petunias, ivy-leaved geraniums and other trailers come into their own. By the time these more substantial plants are throwing a really good show of flowers, the lobelia will be fading away and the gaps will not be noticeable as the fuchsias and petunias will have taken over. Some seed companies illustrate a basket containing nothing but *Lobelia* 'Cascade Mixed'. This is all very well and looks very pretty early in the season but unless very carefully looked after, there is the severe danger of your ending up with a basket full of brown twigs when you really need a ball of colour. You will find some suggestions for using trailing lobelia in the chapter on containers but to be going on with try *Petunia* 'Recoverer White', *Helichrysum petiolatum, Lobelia* 'Blue Cascade', which is pale sky blue, and a white tuberous begonia in the top.

## Lonas

A pretty and exceptionally long-flowering hardy annual which is good in borders and as an everlasting. The heads of flowers, which can be up to 5in (12.5cm) across, are rather like bright, butter-yellow ageratums. The plants grow to about 18in (45cm) and the foliage is dark and very deeply and finely divided. It is generally self-supporting and thrives in open situations in soil that is well-drained and fertile but not rich — perhaps the easiest of all conditions to provide. From a March or April sowing, flowering should last from July to September and if you want to cut for drying wait until the heads are fully open. The flowers are a bright, clear shade well set off by the dark foliage. Dead-heading is beneficial.

**Lunaria
(Honesty)**

Also known as moonwort and satin flower the one widely grown is a biennial perversely called *Lunaria annua*. Essentially a plant for the wild garden where their strong purple flowers and then their silvery flat seed pods are rarely offensive, they can also be lined out in a row to provide heads of drying seed. Few people grow them in a more formal fashion in spite of the fact that if well grown and given the space, they make good bushy plants providing a colour and a spring option not found in other plants. Its delightful scent is an extra bonus.

There are four varieties available. The ordinary simple purple is a fine plant and is the one you will usually find; it is good for everything. Then there is a variety called 'Munstead Purple' with darker, more velvety flowers originally selected by Gertrude Jekyll. And there is white, just as good as an occasional interloper in the border with even less likelihood of a colour clash than the purple and good in bedding too. There are a number of reddish forms and forms with flecked flowers that turn up in the mixture which is sometimes available. Sadly, as if that was not quite sufficient, some meddlesome streak in the mind of man caused a variegated strain to be selected and perpetuated. It plays tricks too. Having had the misfortune to be forced to grow some, after being presented with an unsolicited packet by a well-meaning, but totally misguided acquaintance, you sow it. Up come the plants, almost every seed will germinate, and up come green umblemished plants. The plants get bigger and just as you have found a spot into which to transplant your little treasures they get their own back by suddenly breaking out in creamy blotches! So beware.

All honestys prefer some shade and soil which is fairly well-drained but will be quite happy out in the open, especially if surrounded by other plants. A well-prepared soil helps too. Thin early and wide apart in the seed rows to get good-sized plants for putting out, and try them with a yellow tulip like 'Golden Appledorn' or try the white form with the old favourite 'Queen of the Night' tulip.

**Lupinus
(Lupins)**

There is quite a range of both annual and perennial lupins that fit into our category but sadly, few of either are actually used though they are worth trying in bedding terms.

The annuals are no trouble: 12–18in (30–45cm) high they derive from a variety of wild species from Southern

Europe, Asia and the Americas and seem to thrive especially well on soils which are dry and not particularly fertile — perhaps this really is the annual whose requirements fit the old wives' tale about poor soil. Give them the usual hardy annual treatment and they will be quite happy but thinning is especially important, leading to branched growth and a good succession of flower spikes. 'Pixy' is a mixture of pinks, blues, lavenders and white including some bicolours and grows to about 12in (30cm). 'Yellow Javelin' is twice as big and has tall spikes in bright butter yellow. *L. densiflorus* and its varieties have rather pretty, slightly downy foliage and the flowers are in whorls up the flower stem giving a candelabra like effect. The colours are the most varied of all and include a strong red and a red and white bicolour. All make good cut flowers.

Now the perennials. They can be had in flower in their first season if sown early in warmth and pricked out into pots. Treating them as biennials is risky as they do not move well except when small; by damaging their fleshy roots, disease is encouraged. The dwarf 'Lulu' and 'Band of Nobles', an improved Russell selection, are the ones to try.

# M

**Malcomia (Virginia Stock)**

How this little annual from the Mediterranean comes by its common name I cannot say; it is the victim of the same quirk that consigns *Scilla peruviana* to South America when it grows wild in Italy, Spain and Portugal.

This plant has a strange status in the gardener's mind. At one time separate colours including white, carmine and yellow as well as pinks and lilacs were available, and the plant was much valued. Then its reputation went into a decline and it was thought fit only for young children who appreciated the speediness with which it flowered or to be given away free with gardening magazines because seed companies are able to buy the seed for next to nothing. This is a rare case where Social Democracy prevails for the truth is in between. It is a charmingly pretty little plant, slightly scented, easy to grow, quick flowering and one of the cheapest to buy which will brighten any tiny spot where you care to toss a few seeds. Strangely, for a plant that rarely gets taller than 8in (20cm) and has relatively good-sized flowers, if it were smaller it would be even more useful. It follows, then, that it is at its best when sown in the cracks in paving which is bedded on nothing richer than soft sand, for this keeps it dwarf. In such situations it will seed itself and although there will be a tendency for its colour to drift towards a thick pinkish lilac it makes a happy addition to the limited paving flora.

Do not, please, be tempted by that tired old notion of mixing Virginia and night-scented stocks together so that the flowers of the Virginia stock will make a pretty display while the less appealing flowers of its night-scented relation provide the scent. What always seems to happen is that the night-scented version grows slightly more quickly, and taller too, tending to smother its showier friend which is not clearly seen and tends to be stunted to boot. What is more, the sweet though more delicate scent of the cheerier plant is lost. Other spots for the Virginia stock are the corners of gravel drives and gravel paths, window boxes and tubs where something of more choice has unaccountably vanished, and as a quick flowering edging to paths.

But won't someone spend a year or two re-selecting for separate colours? The seed companies may not make a fortune out of them but it would give the discerning gardener a little more scope for the imagination. And even if they were never sold separately the colours could be blended into a more balanced mixture.

**Malope**

Malope are hardy annuals related to the hollyhocks and harbouring the same dreaded rust disease. Only *M. trifida*, so-called because the leaves have three long, pointed lobes, is cultivated and this is a tough plant, growing to 2–3ft (60–90cm) with erect, though branching, stems which make substantial plants. The flowers are a rich reddish purple, up to 3in (7.5cm) across and are carried in the leaf joints all the way up the stems. They are long-flowering, from June to September if sown in late March, and are excellent cut flowers too. Contrary to what is sometimes suggested, they need a fairly fertile soil which does not get too dry — very light soils are rarely satisfactory unless improved with humus. Sunshine is a great boon to a prolific display. The varieties available are a silky purple and a mixture of red, purple, pink and white.

Finding companions is tricky — if you want to make sure passers-by never peep into your front room, plant a border of the purple malope, *Geranium* 'Picasso' or 'Merlin' and *Petunia* 'Sugar Daddy'. They are sufficiently close in shade for their differences really to jar. In wilder, less formal situations pop odd plants in mixed borders amongst whites and silvers.

## Matthiola (Stocks)

As the years roll by gardeners change in their enthusiasm for different plants. The stocks are a group of plants which are generally grown a great deal less than they once were. Although they are still popular as cut flowers, few people grow them at home for this purpose and prefer to buy them from the florist or market where they are often very good value. Part of the problem for the home gardener is understanding which of the different groups is suited to which purpose — it can be confusing, even to the experienced gardener. And the other problem is the difficulty in ensuring that all the plants turn out to be doubles. Even though the procedure is now fairly well established, it still causes problems for the home gardener.

First, let us deal with the night-scented stock (*Matthiola bicornis*). This is an unassuming little annual reaching about 12in (30cm) with heavily fragrant lilac flowers which open and give off their scent only in the evening. They are best sown in small patches amongst the geraniums and the scent wafting in the window is quite magical.

Now let us have a look at all the other stocks. They are divided into two basic groups. The annuals consist of those grown for summer bedding and for summer cut flowers both inside and out. 'Ten Week' stocks, the 'Beauty of Nice' strain and the 'Seven Week' strain fit in here together with the 'Mammoth Column' type and 'Pacific Column' type. The biennials include those grown for spring bedding such as the 'Brompton' and 'East Lothian' strains.

*'Ten Week'.* So called, as you might expect, because they flower in ten weeks from sowing. They reach about 12in (30cm) high and are sown in March at a temperature of 60°F (15°C) and grown on at about 50°F (10°C) for planting out in May and flowering in June and July. Usually only a mixture is available. Unfortunately, they tend to burn out later in the summer so are best used as short-term plants with a plan in mind to replace them later on. 'Ten Week' stocks are available in 'selectable' form to ensure that all the plants set out will produce double flowers. The

procedure is simply to reduce the temperature to about 45°F (7°C) after germination and when the seedlings are almost ready for pricking out. Soon two types of seedling will be apparent — those with dark green leaves and those with yellowish green leaves; the difference is quite distinct if the temperature is right. Only the seedlings with yellowish leaves will produce double flowers and only these should be pricked out. Plants with single flowers are not worth growing.

*'Trysonomic Seven Week'.* An even earlier flowering type, quickly producing a single flower spike which is followed by side branches. This too can be selected to achieve only double flowers but this is done in a different way. The seedlings are grown in the normal way and all are pricked out but when the young plants have produced about four leaves you will be able to divide the plants into stronger and weaker seedlings. The weaker plants should be removed and only the stronger plants put out. You get about 60 per cent double.

*'Beauty of Nice'.* This is a much taller strain reaching 1½–2ft (45–60cm). It flowers a little later than the 'Ten Week' type, flowers for longer and the stems are long enough for cutting. They are good bedders too although sometimes need support. This strain has an alternative use. Sown in July and August in the unheated greenhouse they can be potted up and make excellent winter pot plants for the cool greenhouse.

*'Column' Types.* For cut flowers in summer both outside and under glass the column stocks are unsurpassable. They produce just one single, but very impressive, flower spike and are the ones seen in florists. They are usually available as selectable at the seedling stage to produce only doubles. Growing to 2–3ft (60–80cm), they need support inside as well as out. There are two types, 'Excelsior' and 'Pacific'. The 'Excelsior' strain has long been favourite for greenhouse growing and, although few amateurs grow them, they could very usefully be grown in cold houses or well-ventilated polythene tunnels by keen flower arrangers. They are sown between October and January, planted out between January and February to flower from April in a heated greenhouse and from May without heat. It is important that the temperature be kept

below 60°F (15°C) and preferably around 50°F (10°C) or bud initiation will be inhibited and all your effort will be wasted. A slightly acid soil is conducive to the best growth and make sure that the plants never dry out or you will get short spikes. The 'Pacific' strain is less widely available but is better suited to growing outside. Unfortunately, it is not selectable for doubles but will naturally produce about 65 per cent. Sow them directly outside like hardy annuals in March or April depending on the area and season. If you sow as thinly as possible no thinning will be needed and you should have spikes for cutting from June onwards. Well worth trying and sowing successionally to give a longer cropping period.

*'Brompton' Stocks.* Bushy, little scented plants reaching 15–18in (38–45cm) and flowering in May and June if sown the previous summer. In most areas the best routine is as follows. Sow outside in June or July in the same way as you would wallflowers but prick out into 3½in (9cm) pots and overwinter in a cold frame or cold greenhouse. They are susceptible to damage in most winters if planted in the open ground although in favoured parts of the country this is a valid option. If you have space, they can be sown in boxes in the cold frame and this can reduce disease problems. Plant them out in mid-March to flower six weeks later, or pot them and grow them as pot plants. You will usually only find a mixture. Selectable strains are available, selection being made on the basis of seedling colour.

*'East Lothian' Stocks.* Similar to the Brompton type but about half the size and flowering after the Bromptons. They are raised in the same way with the same proviso as to winter hardiness. Improved forms on the way will flower earlier and for longer, making them even more valuable. They are not available as selectable but the percentage of doubles is very high. The 'East Lothians' can also be sown in early spring to follow on from the 'Ten Week' and 'Beauty of Nice' types and their long flowering period is appreciated, although in very warm summers flowering can be delayed.

Recently what seems to have been a dramatic advance in the selection of stocks for double flowers has taken place. For the home gardener the manipulation of the temperature to alter the leaf colour and so reveal the

.doubles can be a tricky business — some may want to grow them without a greenhouse at all, let alone have the facilities to provide two temperature regimes. But from Japan comes an entirely new system. Believe it or not, the seed leaves actually have a small notch in one side if the

*Figure 7.11 'Stockpot' stocks have a unique way of showing which are doubles — the seed leaves have a small notch*

flowers are to be doubles; single-flowered plants have no notch. What is more, as the plants grow older, the notch becomes more noticeable so that you get a second chance. There is no need to change temperatures as this method of selection does not depend on temperature at all.

The first variety of this type to be available is 'Stockpot' which comes in a limited range of red and pinkish shades, plus white, grows about 10in (25cm), and flowers about nine weeks from sowing. It makes a good pot plant or can be used in beds and containers outside. You should get about 60 per cent doubles. When this method is transferred to the taller cut flower types, surely we will see a revival in the fortunes of the stock.

Of course the great attractions of all stocks hardly need emphasis — the wonderful scent and delightful pastel shades. Even though some mixtures may have three different pinks the colours are sufficiently distinct to be appealing and all are such definite yet delicate shades that none can possibly be offensive. As cut flowers they are not difficult but are long-lasting and make good companions with a wide range of other plants. Outside, their use is somewhat restricted by their short flowering season and the fact that they do not overwinter well. So plan for short bursts of scent and blending colours in containers and on path corners where they can be preceded and followed as part of a planned succession of plants for a special spot.

**Mentzelia**

A bit of a mystery under this name but at once more familiar under its old name of *Bartonia aurea*. Although a member of a family (Loasaceae) renowned for its stinging

194

hairs this, the most ornamental of the tribe, has the great good sense to be without them, so making cultivation all the more straightforward. *Mentzelia lindleyi*, as we now

*Figure 7.12*
Mentzelia lindleyi

call it, is a rather bristly, straggly, hardy species reaching up to 2ft (60cm) and needing support if it is not to disgrace itself by collapsing onto its companions. The flowers are rather like those of the showier hypericums, brilliant yellow with a mass of yellow stamens in the centre; they can be up to 2½in (6.5cm) across and although the individual flowers are not long-lived, they come in a constant succession. Sadly they also have the inconsiderate habit of closing up in dull weather (and isn't it always dull when you want to show a special friend round the garden?) but nevertheless they make such a show that they are well worth cultivating in sunny sites and the well-drained but fertile soil, which is the ideal of the assiduous cultivator of annuals. To make up for their steadfast refusal to display their charms on overcast days, they have a sweet scent, so maybe can be forgiven.

**Mimulus (Monkey Flower)**

A few years ago no one would have suggested using mimulus as a bedding plant. They were valued as bright summer-flowering plants for bog gardens and waterside plantings and little else, but they are now one of the successes of modern plant breeding. From tall, rather straggly plants of variable habit and with a dependence on plenty of moisture, there are now four good series of varieties and all are good bedders — although there is still a necessity for soil that does not dry out.

Probably the most generally useful series of varieties, although at the moment only available in two colours, is the 'Malibu' series in orange or yellow. Its flowers are the smallest in the group but are carried in the largest numbers. It is the most tolerant of dryish soils and windy sites and it is also the lowest growing, at about 6in (15cm). 'Malibu Orange' has a few yellow speckles in the throat, 'Malibu Yellow' has similar orange speckles and there are ivory and red versions on the way. You may come across the orange and yellow in a mixture called 'Oranges and Lemons'.

Going up in size a step, there is the 'Calypso' series in a wide range of colours including plain faced types, spotted ones and bicolours — at least eight — but this is probably a little less tough and less floriferous. But the wonderful range of colours makes it very popular. 'Quickstep' is similar except that all the colours are plain faced without speckles. The colours are intense without being hard and, like fibrous begonias, make lovely beds without other plants.

The largest-flowered of all is 'Viva' which also grows to about 12in (30cm) tall and really is spectacular. The flowers are bright yellow with a red blotch on the central

*Figure 7.13* Mimulus *'Viva'*

lower lobe of the flower, four, slightly smaller, more spotty blotches on the other lobes and a red speckled throat.

Apart from their intense colours, fresh foliage and their ability, in the 'Malibus' especially, to recover rapidly from bad weather, these new mimulus have another great advantage. They need only cool temperatures during their propagation period. They will tolerate light frosts after planting if hardened off properly and they can be in flower seven weeks after sowing. The result is that you can sow in early March at 60°F (15°C), reduce the night temperature so that the plants are just frost-free, 36°F (2°C), and have the plants flowering at the end of April. For planting out in early May in flower, a sowing in the first half of April will be quite early enough. What I am afraid you cannot do is sow in January for flowers in March because mimulus need a 13-hour day before they will produce any flower buds so the trick is to sow no earlier than four weeks before the 13-hour days start, hence the suggestion for a first sowing in early March.

As these plants are so quick to flower, they are very useful as emergency gap fillers during the summer to replace others which have unaccountably died on you — it always happens somewhere in the garden — so a packet of mimulus can be very useful.

The only problem with these cheery little flowers is that they need rather particular soil conditions. As new series come out, each seems to be less fussy than the last but they must still have a soil which does not dry out. This can be managed by adding plenty of organic matter to the soil before planting, by watering well in even slightly dry spells and by siting the plants in the shade where possible. And indeed they are very good shade plants.

A dry spell in the garden will soon stop their flowering but that need not be the end of the display for the season. The plants can be cut back hard with a pair of shears, watered well maybe with a liquid feed and this will soon encourage more growth and flowers although the second flush is not always as good as the first. Being the most vigorous, 'Viva' seems to be the best in its second burst.

In the garden try mimulus in shady beds with plenty of humus in the soil — dry beds under trees will not be successful. The colours, with their brilliant yellows and oranges, are more sparkling than impatiens or begonias and would look well in front of a tall intense impatiens such as 'Blitz Orange'. 'Viva' could go behind 'Oranges and

Lemons' and red foliage is a good companion too. Plant them in drifts amongst spring-flowering border plants and shrubs, in those odd shady corners by the patio, in containers, in dark parts of the garden which need brightening, and of course they are ideal for window boxes on the north side of the house.

**Mirabilis (Marvel of Peru)**

And a marvel it is too. For one thing the flowers only open in the late afternoon. During the day the buds fold so tight the plant looks really rather unremarkable, not the sort of thing that anyone would normally suggest growing. But in the evening, when the flowers open, the plant undergoes a transformation. The flowers are 2in (5cm) long with narrow tubes opening to a flat lobed face. The commonest colour is a strong magenta but a bold yellow is also common and white and pink will sometimes turn up. The strangest thing is that every now and again a whole branch of the plant will suddenly produce flowers of a different colour. Yellow flowers appearing on an otherwise magenta plant seems the most regular variation. Spotted, streaked and speckled flowers also often appear and some flowers will be half magenta and half yellow. On top of this curiosity there is the scent — which wafts hauntingly through the still evening air.

Although usually treated as a half-hardy annual, and raised in the normal way, it is actually a perennial plant, forming tubers rather like a dahlia. In warmer areas it overwinters happily and in Corfu it is common in gardens where it forms large plants up to 3–4ft (0.9m–1.2m) high. These tubers can be lifted and dried in the same way as dahlias, making sure they are not frosted in the winter months, and replanted in the spring. Plant them near a place you pass in the evening or by a west-facing patio where you sit out. As they are unprepossessing early in the day to say the least, train one of the less vigorous ipomoeas through them to flower in the morning and close up by the time the mirabilis opens.

**Molucella (Bells of Ireland)**

A curious member of the mint family from Western Asia, this is a plant which has only recently become widely grown. It makes an interesting garden plant but it is for drying that it is usually grown. A half-hardy annual reaching about 3ft (90cm) with coarsely toothed leaves it

198

has tiny pink flowers in whorls of six or eight, each flower cupped in a shell-like, heavily veined cup. It is the succession of these up the stem which, when dried, gives their appeal. Grow them like any other half-hardy annual, though a temperature of 60°F (15°C) is enough for germination and a relatively cool temperature suits them best when growing on. They can be sown outside in late April too.

In the garden they make interesting random additions to the mixed border and will seed themselves enough to keep appearing from year to year, unless you use the hoe — which should be banished from the real enthusiast's garden.

**Myosotis (Forget-me-nots)**

Another great British plant, in the wild, in cottagey and informal gardens and borders, in formal bedding and as a pot plant. There seems to be some doubt as to whether the varieties we find in catalogues are developed from *M. alpestris*, a rare British native that grows on alkaline rocks in high places such as Ben Lawers in Scotland and the Cumbrian fells, or from *M. sylvestris*, a far more common plant which those who remember their school Latin will realise grows in woods, especially damp woods and is far more common. The consensus seems to be that the very dwarf types such as the 6in (15cm) 'Blue Ball' are derived from the alpine type while the taller ones such as 'Blue Bird' are developments of the woodland form.

Although both are perennial in the wild, the cultivated forms are usually grown as biennials and rarely persist — often because they are killed by mildew. They do, however, self-sow. In informal situations, it is the taller varieties that are the most effective. The little blue buns that are suited to containers, window boxes and small-scale bedding schemes just do not look right at all. Taller, slightly looser growing types such as 'Blue Bouquet' at about 15in (38cm) and 'Blue Bird' at up to 2ft (60cm) seem to blend with the tumbling approach of organised chaos that makes cottage gardens so delightful. They will seed themselves and, providing they do not cross with any wild cousins in local woods, flower size should stay quite large. They can then be left to grow if they fall in the right spot and removed, or maybe transplanted, if they do not. These tall ones are sometimes also the best for growing under or in front of very tall tulips in spring bedding schemes. Smaller varieties tend to leave such a long piece of tulip stem exposed that the result looks rather silly. In more formal situations the smaller varieties are usually more effective. In combination with other spring biennials, especially wallflowers, they really are excellent. Shorter tulips work best and the developments of *T. fosteriana, T. griegii* and *T. kauffmanniana* will usually be my first choice as associates.

The one great problem with forget-me-nots is mildew. And it is a powdery mildew that attacks them which is especially bad in hot conditions, like the mildew that attacks roses. The gardener is therefore at the mercy of the weather and so spraying is the only answer if the tell-tale white felting starts to appear. Fortunately, the end of the spray that is used on the roses will do a good job and, of course, the weather that is conducive to this mildew does not usually arrive in Britain, if it arrives at all, until the flowering period is over. It does though make the plants very unsightly and can disrupt seed production, which is a shame if you are hoping they will naturalise in borders.

They are raised by the standard biennial treatment and should be sown in June. In especially wet areas it often happens that a number of plants rot off during the winter leaving a distressingly gappy display. In most areas the answer is probably to pot up a few and overwinter them in a cold frame which is protected from the worst of the winter wet. Gaps can then be filled in the spring. The slightly hairy leaves and the dense basal branching encourage water to collect in the centre of the plants and this is doubtless the cause of the problem.

Myositis come in a number of blue shades although there is not a very dark one. There are also some pinkish shades such as 'Carmine King' and 'Rose Pink' but they seem to grow a little taller than most of the dwarf blues and so make rather a rolling carpet. There is a white too, which somehow does not quite come off. 'Royal Blue' is probably the deepest colour although the plants reach 12in (30cm) and so are not ideal for all situations.

# N

**Nemesia**

A brilliant and easy plant with a dramatic failing — but a failing which can be accommodated in the garden with a little thought. Only one species is generally grown, *N.*

*strumosa*, but the cultivated forms are shorter than the 2ft (60cm) it reaches in its African home and now have a brilliant and comprehensive range of exhilarating shades. Sown in the normal way and given the usual half-hardy treatment they will usually be in flower when they are put out in late May. Try to keep them growing steadily because a check to growth can be surprisingly detrimental. They do not need pinching, being one of the most naturally well-branched of all annuals, and romp away in all but the most inclement seasons to produce quickly their vibrantly coloured flowers. Nemesias have a preference for strongly fertile soil and one which does not get too dry; if it is on the acid side they will be especially happy and sunshine is usually welcome.

They flower quickly and make a good display before other summer plants are really into their stride and, sadly, this leads to their one drawback — they tend to give over rather early. So it is no use making a fat group of them at the front of the border unless you have something in mind with which to follow them. More nemesias is sometimes the answer and another sowing in May in an unheated greenhouse will provide the plants which can be set out from pots to replace the early ones. Alternatively they can be used, especially in containers, to give an early show and simply left in place for other plants to grow over. They make excellent and very long-lasting cut flowers too although tending to pale rather if not in good light, so that the intensity of colour is lost in the flowers which open after arranging. In cool summers like that of 1985 they will flower for far longer than in the hot dry summers like 1984 when they burnt out very early.

As pot plants they work well and provided the greenhouse is heated to around 45°F (7°C) they can be had in flower all the year round. The very shortest varieties are of course best suited to pot growing, 'Carnival' at 7–9in (18–23cm) is probably the best bet with flowers in various reds and bronzes, pinks, orange, yellows and white. Grown outside, this was one of the most impressive of all the bedding plants on trial at two trial grounds in August 1985.

In borders a number of others are worth trying including 'Funfair' which concentrates on the more fiery shades, 'Fire King' which is a vivid scarlet, 'Blue Gem', a pure, dark sky blue with a tiny white eye and 'Tapestry', a relatively new one with possibly the widest range of shades including three different blue and white bicolours. 'Blue Gem' in particular is a much underused plant, not unlike a taller, more upright version of *Lobelia* 'Cambridge Blue'. The flowers may be smaller than those of other varieties but set against the fresh green leaves they are very impressive.

## Nemophila (Californian Bluebell)

Now here is a plant that defies all the old fables about the sort of soil in which annuals do best. They are rather sprawly, with pale, slightly sticky leaves and blue flowers, growing little more than 12in (30cm) high. Their preference is for a moist soil, though not a clay one and shelter from the worst of the sun too. As they are so lax in growth and so incapable of supporting themselves, shelter from the worst of the weather is no bad thing and a little discreet staking with short twiggy brushwood can be a wise precaution. As you might expect they do best in the summers most of us complain about the most — cool and rather damp and in a heatwave they tend to shrivel. They can often be best displayed in large urns or window boxes on north-facing walls where they get no sun and will dry out only slowly and yet will not get too drawn as the light will still be fairly good.

This is another case of a dramatic reduction in the number of variants available over the years. You will usually only find two — *N. menziesii* and *N. maculata*. *N. menziesii* (also known as *N. insignis*) is pale blue drifting to white in the centre, the flowers each being about 1½in

*Figure 7.14*
Nemophila
maculata

(4cm) across although once there were purple, claret and many spotted and streaked types available. This grows to 9–12in (23–30cm) and is a very dainty plant. *N. maculata* is rather smaller and an unusual colour. The flowers are white with slightly purplish veins and a large purple spot on each of the five lobes of the flower.

**Nicandra (Shoo Fly Plant)**

You will doubtless get the message from its common name that this plant keeps flies away. Well, so it is rumoured. Suffice it to say that when I have grown it flies have been no more rare or more common in its vicinity than anywhere else. It is said to repel whitefly, which would indeed be much to its credit but the young plants in my greenhouse were not immune to the attentions of the little beasts. Try it by all means but please do not write rude letters to the seed company if it is totally ineffective. A member of the potato family, *N. physaloides* is the one that is usually grown, and it makes a stout plant about 3ft (90cm) high with handsome (i.e. large) leaves and small pale blue, almost tubular flowers which are not freely produced and which do not open for very long each day. They like a rich fairly moist soil and a sunny situation; the usual half-hardy treatment suits them.

To be honest their main appeal is the fact that you can point them out to visitors and tell them they do not work. If you want to grow something with 'handsome' foliage, grow something really dramatic like ricinus.

**Nicotiana (Tobacco Plant)**

Of all the summer plants that are blessed with a strong perfume, surely the tobacco plant is indispensable. Maybe it is because the scent is so powerful in the evenings, which tend to be restful periods for everyone, that it is so appreciated. The one that has the best scent is not a highly bred cultivated variety but a wild species from South America, *Nicotiana alata*. True, the one usually grown is a slightly larger-flowered form (sometimes called *N. affinis*) but it is nevertheless a very simple plant. Reaching 4ft (1.2m) in height and carrying loose heads of large, slightly creamy white flowers, you can feel the scent wafting towards you as you step into the garden on still evenings and it is so persistent that it drifts for quite a distance. The plants are big and bushy, breaking well from low down if given the space and if pricked out into 3in (7.5cm) pots and spaced out on the greenhouse bench. Unfortunately they can be rather prone to aphids. Give them the usual half-hardy treatment and to get the best germination leave the seed uncovered. One interesting thing I have discovered is that they persist well from year to year and they do this in two ways.

First of all they are perennial plants and even after the winter of 1984/5 roots left in the border, sheltered only by a fence, survived. They were over 2ft (60cm) high and in flower from the old roots by the first week in July. The same was true in previous years of more dwarf strains but in 1984/5 only *N. alata* overwintered. Other types nevertheless self-sowed and, by May, so many seedlings were popping up that many had to be removed. A number of plants in a variety of heights and shades had started to flower at the end of July and a tallish dark red one looked very nice amongst the roses.

Apart from the tall *N. alata* there are three strains widely grown, the 'Nickis', the 'Dominoes' and the 'Sensations'. The 'Nickis' come from America, are 15–18in (38–45cm) tall and come in six shades. Usually only the mixture is available but single colours can sometimes be found in catalogues. They are very prolific and flower for many weeks but the scent is not as strong as it might be; the usual half-hardy treatment does them quite well. In 1979 the various colours took half of the awards given in the RHS trial. Since then another group, the 'Dominoes', has been developed in Britain and in an interesting way.

We all know that in good seasons petunias can produce the most dramatic of British summer displays but in poorer years they do not perform the way they do in California, where so many of the varieties have been bred. Various attempts have been made to improve the resistance to poor weather in petunias but the old 'Resisto' series is still the best. So another approach was tried. Nicotianas, which are closely related to petunias, cope far better with the eccentricities of the British weather so Mike Hough, of the plant breeding company Floranova, decided to work on them. Rather than try to improve the bad weather resistance in petunias, he would attempt to boost the flowering capacities in nicotiana and at the same time improve their tolerance of the hot, dry conditions which suit petunias.

So there were three main aims in improving nicotianas: to maintain and if possible improve their resistance to wet and windy weather, to improve their tolerance to hot sun and drying winds and to improve their flowering capacities. The particular improvements to their flowering qualities that were required were early flowering, bushiness of growth through increased branching from the base of the plant, increased range of flower colours, increased floriferousness and reduced height.

This task got under way in 1979 and as early as 1982 the first varieties were introduced. The speed with which this was achieved was in itself remarkable and was the result of using all the latest plant breeding techniques. The 'Nicki' strain, which was the most widely grown variety at the time work started, was used initially but being rather tall another, more dwarf variety 'Idol' was also used and as a result of various crosses and selections the 'Domino' strain was created.

Increased tolerance to hot sun and dry winds was achieved in an interesting way. In many varieties there is a tendency for the petals to be rather shiny and the wind and heat quickly draws water from the petals which then soon go limp and collapse. By careful selection a slightly heavier and more waxy flower was developed that did not lose water in the same way. The height was reduced to 12in (30cm), although in rich moist soils they may grow a little taller, and to make really bushy plants the degree of branching low down on the plant was increased.

A significant improvement was also made in flowering capacity. This was done in an unusual way. Normally the flowers of the nicotiana are carried so that the flat face of the flower is held at right angles to the tube of the flower behind it, but early on in the work it was noticed that some flowers carried the faces at a different angle, they faced upwards at 45 degrees instead of outwards and as the gardener looks down, this gives a much clearer view of the face of the flower — in short the plant looks far more colourful. This characteristic was bred in with the others and is a unique feature of the 'Domino' varieties.

In 1982 four colours were made available — red, white, crimson and pink with a white eye. Soon four more colours appeared — lime green, purple, picotee and purple with a white eye. Soon it is hoped to add the last two shades — rose pink and rose with a white eye. Many of the colours in nicotiana tend to be rather dowdy so a conscious decision was made to add the varieties with white eye to lighten and brighten the display when the mixture is grown.

Most catalogues list 'Domino Mixed' and some list a selection of the separate colours too. They are ideal plants for small gardens and small beds, at the front of beds in larger gardens and they are compact enough to be used in containers such as tubs and window boxes. The colours are excellent and the white in particular is one of the best of white bedding plants, their eye colours are bright and the crimson is strong but not harsh. The 'Dominoes' are neat of course and not subject to the battering by wind and rain. Some of the old varieties do tend to close up during the day but the 'Dominoes' are open all day.

'Sensation' is a variety older than both 'Nicki' and 'Domino' and much more of a cottagey plant. It grows to about 4ft (1.2m), the flowers stay open all day and they come in a lovely range of shades including the best purple and the best dark red in any strain. There are also a number of very soft shades in which the backs of the petals are dark — the cream with apricot petal backs is outstanding.

One odd variety that does not fit into any of these series is known simply and descriptively as 'Lime Green' and the name adequately describes the colour of the flowers. The plants reach about 2ft (60cm), are very prolific and weather resistant and make excellent and interesting cut flowers as well as unusual border plants. The colour is not sufficiently piercing to clash with the majority of other shades so you can be fairly free in its use.

All nicotianas like a rich and fairly well-drained soil with plenty of sunshine and even the tallest varieties rarely need staking unless they are planted in very exposed sites. Although they will self-sow it is still better to sow under glass rather than direct into the open ground as they are martyrs to slugs.

**Nigella (Love in a Mist)**

These surprising members of the buttercup family are hardy annuals from the Mediterranean, and are often seen in the Greek islands where they vary slightly from island to island. They were once cultivated for their seeds and were also an occasional agricultural weed in Britain. At first sight you would not put them with buttercups or even delphiniums. The plants are stiff and upright, the foliage fine and open, and although the single-flowered types have the five parts of a buttercup, it is the curved tips to the immature seed pods that give them an unusual appearance.

There seem to be only three varieties easily available. The semi-double 'Miss Jekyll' is a lovely cut flower in sky blue which has been around for many years. It grows to

*Figure 7.15* Nigella *'Miss Jekyll'*

about 18in (45cm) with long stems and is ideal for cottagey arrangements. It can also be grown in the border where, if dead-headed, it flowers for weeks. 'Persian Jewels' is a little shorter and includes two lavender blue shades, a couple of pinks, and a not quite pure white. It too is good in annual borders and for cutting.

A species from Spain and North Africa has recently re-appeared in catalogues, *N. hispanica*. It grows a little less than 2ft (60cm) high, has larger flowers than the other two varieties at about 2½in (6.5cm) and comes in a lovely deep blue-lavender. The appeal is much increased by the bright orange anthers which create a lovely picture. It is probably not as floriferous as the previous varieties but the individual flowers are so unusual that it is still worth growing.

All these varieties have very attractive swollen seed pods after flowering and these are often dried for indoor arrangements. Leave them to dry naturally on the plants and cut them as the capsules open to shed the seed. Do not strip off the dried foliage, the stiff wispy strands add to the appeal.

Nigella is a perfectly tough hardy annual for sowing in spring or in autumn in well-drained soil. If, as you should, you keep your cold greenhouse operational all winter, a few plants in pots from a September sowing can be overwintered and planted out in March for some lovely early cut flowers.

# O

**Oenothera (Evening Primrose)**

Almost all the species in cultivation are perennials but many are quite happy treated as half-hardy annuals or as biennials. Although commonly known as evening primroses, with the implication that they open only in the evening, there are a number of good day-long flowering plants which belie this suggestion. Those treated as half-hardy annuals are generally small plants often with disproportionately large flowers for sunny well-drained sites. The flowers are large fragile cups in yellow or white facing directly upwards as the proud gardener looks down to admire his or her handiwork.

*O. caespitosa* has pure white flowers up to 3in (7.5cm) across on grey green foliage which reaches little more than 9in (23cm) in height. Sow early, prick out into pots and plant in full sun and a well-drained soil for flowers all summer. *O. albicaulis* is similar although the flowers fade to pink before collapsing. *O. triloba* is a true hardy annual with small, bright yellow flowers all summer on plants just 4in (10cm) high. *O. missouriensis* is a popular perennial which can be raised from seed to flower in its first year. It has big lemony flowers all summer and reaches no more than 9in (23cm).

There is a great deal of confusion about the names of the taller biennial types. So sticking to the names under which they actually appear in catalogues, *O. biennis* and *O. erythrosepala* (which I can say is the same as *O. lamarckiana*) are the best. *O. biennis* reaches about 4ft (1.2m) and is best sown in June to flower the following season from July onwards. It is a stout plant, of upright growth, with slightly sticky foliage which is very susceptible to greenfly. It was once cultivated for its edible roots. The flowers are pale yellow on long leafy spikes. *O. erythrosepala* is a little taller, with larger flowers up to 3½in (9cm) across.

The smaller more creeping types are ideal for the front of sunny borders, and on the rock garden will survive happily from year to year if the soil is as well-drained as it should be. The taller types are best in mixed borders, flowering a little too late to be treated as true bedding plants. They can seed themselves rather too easily for comfort so remove the dead flower stems, although you will probably find that the first seed capsules are ripe while there are still buds to open.

**Onopordon (Cotton Thistle)**

Now here is a plant you can have some fun with! *O. arabicum* is a huge biennial plant, easy to raise from seed, which in its first season produces broad rosettes of slightly spiny, heavily felted, white leaves. It looks very innocent and very pretty as a seedling in a 3in (7.5cm) pot, which is the stage at which you give it to friends. During the remainder of its first season after planting it can spread to

*Figure 7.16 Raise onopordons in pots and plant them out before the roots become too restricted*

as much as 2ft (60cm) across which people sometimes find inconvenient and it is only fair to warn them that this is likely to happen. Mind you, it looks so mild mannered in its little pot that they will not believe you. But it is about May the following year when they start ringing up with an edge in their voice and by the end of June they are usually frantic! For in the spring it starts to grow. And then it grows some more, sending up its white and woolly flowering stem, and then it just carries on growing to make a huge candelabra of grey winged stems about 8ft (2.4m) high. When it approaches this height it starts to fall over, or rather the side branches collapse into the surrounding vegetation — or what is left of it after the smothering foliage has done its work.

In 1985, in the middle of July, my own plant dominated the border. It had collapsed through the 'Iceberg' rose on one side and through the soft pink 'Mary Rose' on the other and the flowers were peeping through the stems. Some seedlings of the purple-leaved version of *Atriplex hortensis* somehow struggled through amongst the branches and some flowers of *Phygelia capensis*, which seems to be on its travels up and down the border, also added to the performance. A plant of the strong blue *Delphinium* 'Blue Bird' had almost given up, but a few flowers peeped through. And the onopordon looked wonderful, although once it gets loose at the roots, the whole monstrous edifice will start to heave and then will be the time to have it out. This will leave a huge gap but as long as the border gets a drink the roses and phygelius will soon fill the gap and this year's plants — already being raised by sowing seeds individually in 3in (7.5cm) pots in a cold frame or cold greenhouse — will go elsewhere. And to different friends.

# P

**Papaver (Poppy)**

The vibrant scarlet poppies, once so common as cornfield weeds, give some idea of the brilliance and intensity of colour that poppies have and many of the best varieties capitalise on this characteristic. There are also softer shades, with the same delicacy of form and these are as effective in their own way. There are two groups, the annuals and the biennials although some of the latter are short-lived perennials which are best renewed every year.

There are two good varieties based on the red field poppy, the 'Reverend Wilkes Strain' and the 'Shirley Double Strain'. The good clergyman's strain is single and semi-double with a wide range of colours but all in the red and pink range plus white; it includes a good number of picotees and bicolours. The 'Shirley' strain is double-flowered in much the same range of colours, although it does rather depend on the source. In theory it should have no reds. Some strains of 'Shirley' are in fact almost all single-flowered; you take your chance I am afraid.

The other annual species that provides some lovely varieties is *P. somniferum*, the opium poppy. Those we grow are indeed relatives of the plant that has brought so much relief, and so much suffering, to the world but in Britain of course it is never grown for its opium — it would be illegal if it was. Nevertheless I once saw, from a train, a whole allotment planted with lilac opium poppies; I have no idea why they were there. The foliage is rather more interesting than that of the previous group, being much more substantial and a clean sea green in colour. Most of the varieties now grown are double apart from 'Danebrog' whose flowers are shimmering red with a big pure white cross at the base of the petals and with the tips of the petals finely fringed; altogether very impressive. 'Lady Bird' is a black crossed equivalent but without the fringes. The others are double and usually referred to as paeony-flowered, from the extreme doubleness; they can be up to 4in (10cm) across and come in a range of pastel pinks and lilacs. There is also 'Pink Chiffon' available as a separate colour. For dried arrangements where the seed pods are very useful, you sometimes come across a variety with especially big pods grown specifically for this purpose. All these varieties are perfectly hardy and indeed do not usually thrive if treated as half-hardies and transplanted. So give the usual hardy annual treatment being especially ruthless with the thinning. They like sunshine, well-drained soil and shelter from the worst of the wind which can easily damage the plants and the delicate flowers.

The biennials, or those which can be treated as such, are varieties of the 'Iceland Poppy', *P. nudicaule* or the alpine poppy *P. alpinum*. 'Champagne Bubbles' is becoming the standard Iceland poppy and this is a recent variety reaching 18in (45cm) with a wide range of colours tending towards the pastels and including red, bronze, apricot, pink and yellow. They are superb for cutting, as are all of this group, but they must be picked as the buds are still unfurling and the bases of the stems dipped at once into boiling water to seal them. 'Sparkling Bubbles' is similar but in a more intense range of colours — and also a little cheaper. 'Matador' is a single-coloured variety in brilliant scarlet which is very impressive and there is also a relatively recent introduction with an amazing range of colours. These are the 'Oregon Rainbows'. It took 20 years to create this strain which was bred in Oregon as a cut flower. The flowers are big, the first one may be exceptional at 6in (15cm) across, but they will then settle down to 4in (10cm), and they will be carried on stems about 20in (50cm) high. The range of shades is astonishing with fewer of the familiar reds and oranges but far more other colours. These include many pinks, apricots, creams, lavenders, lemons and white plus green, picotees, bicolours and some will grade gently from one shade to another. The combinations are almost endless and there are semi-doubles as well as singles.

Iceland poppies can be sown in January if you want them to flower the same summer but the best way to get really good plants is to sow in a cold greenhouse in summer and prick out into 3in (7.5cm) pots. The plants then go into the garden in September to flower in May the following year. They can be sown outside, thinned and transplanted but poppies often object to disturbance and will die on you, so raising them in pots is usually the best way. Again, sun and a well-drained soil are essential and the more you pick the more they will keep flowering.

Rather smaller than these cut flower types is 'Garden Gnome', which is only a baby at about 12in (30cm). You can cut the flowers for the house, but it is really as a garden plant that it is valued. The stems are absolutely upright and come in great continuous profusion. The colours are an orange just a little short of red, yellow, pink, salmon plus a pure white.

The alpine poppy, *P. alpinum*, just seems to come in mixed, which does not sound very exciting but do not be fooled. From little tussocks of foliage rise short stiff stems about 6in (15cm) high with delightful miniature poppy flowers. Mind you, although they may be smaller than the Iceland types, they still make a real blaze in orange and yellow shades in particular but also include white, pink, apricot and scarlet. It is vital to take off every seed head the minute the petals fall if you want the display to last otherwise you will have plenty of seedlings for next year but a much curtailed flowering display this time. Ideal trough plants, they stay compact and neat and the open soil in most troughs suits them very well. Plenty of sun is again required.

**Parsley (Petroselinum crispum)**

'Why are you growing so much parsley?' the conversation runs. 'For the summer bedding, the answer follows. Which in turn is followed by astonishment. I must say that parsley is not the most widely grown of bedding plants. It does not flower, or at least you hope it does not, and its leaves are green — not much of a recommendation you would think. But parsley is one of the most useful plants for temporary summer schemes.

There are three approaches. It can be used for the inherent value of its crispy, curled foliage and its intense vivid colour; it can be used as a carpet under more colourful plants; or it can be used simply to plug gaps where you have run out of petunias or where the odd plant gets sat on by the cat and never really recovers. The best variety for bedding is 'Bravour'; it is the most intensely *green* green and in a mat makes a gently undulating carpet — it is also most reluctant to run to seed. And of course picking the odd leaf for the kitchen is not going to affect the display dramatically. I first saw it used by Brian Halliewell at Kew many years ago with yellow petunias and this pair also goes well with lime green nicotianas — either in a container, small bed or in a gap in a larger bed. Other plants with which it goes include almost anything strongly white like petunias or nicotianas, yellows and especially pale yellows like *Pyrethrum* 'Golden Moss' or *Helichrysum petiolatum* 'Limelight'. It is worth trying as a background to specimen plants like standard fuchsias planted outside for the summer and any corner amongst shrubs, perennials or annuals can be filled with a few parsley plants to prevent the weeds getting a hold and to make a background against which other plants can be shown off.

It is best raised like any other half-hardy annual, if it is not to be grown in the vegetable garden, with two provisos; it should be grown in pots and also fairly cool after germination otherwise it gets tall and sparse and takes a long time after planting to make an attractive carpet. There is also what is generally thought of as the germination problem: the seed does take quite a while to come up — it has been suggested that boiling water poured over the seed before sowing helps, but I have found that as long as a well-drained compost is used, and you do not keep it too wet, it is just a matter of waiting and once they peep through things usually go quite well.

Any soil suits parsley as long as it is not too gravelly and sun or shade will be equally successful although, as you might expect, it gets a little drawn in the shade. The crux of the matter is that the soil should not get too dry.

**Penstemon**

These close relations of antirrhinums and foxgloves are unreasonably neglected. Although often bought as plants from specialists in border perennials, and treated as slightly tender plants with cuttings overwintered frost-free as a safeguard, they are easy to raise from seed. The flowers are larger than antirrhinums and quite open faced and although the colour range is not as wide, there are such lovely bicolours that this makes up for the absence of yellows.

There are four good varieties to choose from and they vary in height from about 14in (35cm) up to 2½ft (75cm). 'Early Bird' is the shortest and comes in a good range of shades. 'Skyline' is probably the best of the bunch reaching 14–18in (40–45cm) with many shades including various pinks, pure white, some good dark reds and plenty of bicolours. 'Bouquet' makes 20–24in (50–60cm). In this variety at least 80 per cent of the plants will produce flower shoots on which the flowers are distributed all the way round, instead of hanging down along one side only. It has a good number of blues in its colour blend along with the reds, pinks and violets and is a rather cheaper variety too. Finally, the tallest of all is 'Monarch' also listed sometimes as 'Finest Mixed', reaching 2½ft (75cm). The bicolours in this mixture are especially pretty, some even have two colours in the throat.

Penstemons do not need the high germination temperatures of some plants, 60°F (15°C) is sufficient and the usual half-hardy treatment suits them fine. If it fits into the way you do things, they can be sown quite early, say late January or February, and set out in late April or early May — as long as they are hardened off carefully. They all make lovely cut flowers and are actually perennials albeit slightly tender, and they may well survive the winter outside.

**Petunia**

The petunia has undergone a dramatic transformation. Breeding has brought about a change almost as great as the change from the wild rose to the Hybrid Tea. There are now single and double-flowered types, picotee-flowered varieties, striped varieties and the colour range and the size of the flowers are exceptional. They have come a long way from the wild species from South America with their small white, purple or magenta flowers and their straggly growth.

Petunias are generally divided into two main groups — the multifloras and grandifloras. And these two names fairly accurately sum up the main differences. The grandifloras tend to produce the largest flowers and in ideal conditions make the best show. The sepals, the bud segments behind the flower, are large and rounded. The multifloras are smaller, there are more flowers and they are produced more frequently; their sepals are small and pointed. After damage from bad weather they tend to recover well. They are also more resistant to damage from rain in the first place. The plants tend to be bushier too, and less straggly.

Grandifloras and multifloras are used in very different ways in the garden. For bedding out, the multifloras, and especially the 'Dwarf Resisto' varieties, are most certainly the best. The carpet of colour will be continuous, even in cooler summers which petunias generally do not like although you will find the blue, and especially the pink, are the best. They much prefer dry seasons with plenty of sunshine. So try and find a place for them in full sun and bear in mind that they will put up with a fairly dry soil once they get established although they do not like to be baked. A moist shady situation will only encourage lush growth and few flowers.

Try and go for fairly small plants if you are buying them from a garden centre. Large pot-grown plants seem to be reluctant to send roots out from the root ball and also tend to get rather bare at the base by mid- and late summer. The impact of the display is then lost. Go for smaller plants which may not even be in flower. Of course if you are growing them yourself, by sowing them during the second week in April you should have plants of about the right size.

It can be a good plan to buy grandiflora types in pots, especially as they flower early which is important if you want a really long show. But pot-grown plants in flower are, of course, more expensive. Being that little bit more vigorous they seem to grow away better after planting but also tend to get slightly long and straggly later. If you want to plant them in the open garden I would suggest trying to find small plants or growing your own. Otherwise by August they will be sufficiently tall to be knocked down by

215

a heavy shower — and no one wants to have to stake petunias. Once beaten down you will find an unappealing bare patch of naked stems in the middle of the clump.

In baskets, tubs and window boxes they are super but make sure you plant something else to cover up all those bare stems later on. To keep the grandifloras flowering well dead-heading is important. Petunias are not the easiest plants in the world to decapitate although some varieties, like 'Sugar Daddy', have flowers which snap off quite easily when they are finished. Others have flowers whose stems are soft and stringy and difficult to pinch out. But only with plants in tubs and other containers is it really worth the time.

Raising petunias from seed is not as easy as raising some other plants. The method outlined for begonias on page 93 suits them and they need fairly high temperatures to start them off and the seed is very small — as well as being expensive. But there is another way.

Petunias have definite perennial tendencies and in Britain the frost kills them, but they can be increased by cuttings. This method was once used to increase the doubles before good reliable seed-raised types were available. One spring, in the last week of April, I found an adventurous local nurseryman selling pot-grown grandifloras in flower. Now I would not normally buy them until the middle of May but I bought one pot, for less than the cost of a packet of seed, and was able to cut the plant up into ten cuttings. I picked an especially tall and straggly specimen with a number of stems and cut it into pieces, each with three leaf joints. A cut just below the bottom leaf and above the top one tidied them up and the leaves were removed from the base and flowers and buds pinched out. They were dipped in rooting powder, though I am not sure that was really necessary, and split between two 3in (7.5cm) pots of compost. The compost was made up of half and half peat/perlite. They just sat on the greenhouse bench amongst the seedlings and although they went a little yellow at first, in two weeks, at a minimum temperature of 45°F (7°C) they were rooted. They were then potted individually into 3in (7.5cm) pots and by the time I planted them out at the end of May they were in full flower. It is a good reliable way of saving a little money.

Every year more and more varieties appear and it is sometimes hard to know which to pick. If I were to pick just one it would be a multiflora, 'Dwarf Resisto', which comes in a mixture of at least eight colours. The blue, and especially the rose, are noticeably good, and 'Resisto Rose' must be one of the great bedding plants of all time. You will find that the rose pink shades in other varieties tend to be the most weather resistant too.

There are two rather more recent series of multiflora petunias which are also especially good. The 'Prio' series was bred in Holland with northern European conditions very much in mind and comes with relatively large 3½in (9cm) flowers in five colours and a mixture. The other, newer series, 'Pearl' is different in that the breeders have aimed for smaller flowers, about 1½in (4cm), but have managed things so that they are produced in astonishing profusion. They did very well outside in the dull and wet summer of 1985 and the azure blue variety was one of the most successful petunias of the summer. There are eight colours in the mixture and the plants are rather shorter than those of other varieties.

There is also a mixture called 'Plum Pudding', known in America as 'Plum Crazy', which is made up of heavily veined flowers in various pinks, purples and lilacs plus yellow. This is a quite distinct mixture and a welcome change from some of the more blatantly stunning colours.

Two of the single-coloured varieties are worth growing. 'Orange Bells' is an unusual colour; orange is the best word you can use to describe it but the soft texture of the petals somehow mellows it. It has a white throat and is an excellent hanging basket plant, not getting too lax and being covered with flowers for a long period; it looks well without other plants. Finally amongst the multifloras comes 'Summer Sun'. None of the yellow petunias is all that good, but the genetic weakness associated with yellow is masked to a significant extent in ideal conditions. In Britain where conditions are not ideal, the yellows do not generally thrive but 'Summer Sun' is about the best and can be very effective in good years.

Amongst the grandifloras there are again three outstanding series and a number of single colours. The 'Cloud' series have very large flowers, about 4½in (11.5cm) across, and it is only fair to say that in bad weather they do not stand up all that well. But in sheltered positions, in containers, in porches or on patios protected from the wind, they are superb. There are eight colours, one or two of which are sometimes to be had separately, including two star types and a veined purple shade. The 'Express' series is of more recent origin. For a grandiflora its weather resistance is quite good but it is still nowhere near as good as most of the multiflora types. There are 13 colours, including picotees and stars, which are 4in (10cm) across, but at the time of writing it is only a mixture of ten of the colours (with no picotees) which is generally available.

For picotee types, the 'Picotee' series from Japan is outstanding. There are four colours — red, rose, blue and velvet (purple), all with a distinct and relatively stable white edge. Previous varieties have shown dramatic variations in the patterning.

Even the 'Picotee' series is not entirely stable but the flower marking will tell you what conditions are needed to give their perfect mixture of dark centre and white edge. If the season is wet, the temperature low and the plants are overfed the white picotee edge will be very narrow or non-existent. In very hot, dry seasons and when nutrients are in short supply the white edge will be so broad as to reduce the dark colour to a central star. By adjusting the growing conditions as best you can the ideal markings can be restored.

There are three single colours which are especially worth growing. 'Recoverer White' is probably the best white petunia around and is a wonderful hanging basket plant to grow with silver helichrysum. Its big, pure white flowers last well and recover well from bad weather. I grow it every year. 'Chiffon Magic' is one colour from the 'Magic' series of varieties which have been around for a few years and is the most soft and delicate lilac pink shade. So delicate is it that in full sun it fades to almost white although the slightly creamy eye is always retained. This is the variety for any soft pink/blue/silver schemes and as long as you give it just a little shade it will retain its colour well. Finally another newish· one, from a much larger series. 'Supercascade Lilac' has enormous soft lilac flowers and in the summer of 1985 this was one of the few grandifloras to match the best of the multifloras. My notes describe it as 'unique' at the end of July, 'superb' at the end of August and 'stunning' in the middle of September!

All the varieties so far described are singles and they are by far the most common. Double-flowered forms do appear in most catalogues but their performance in the garden is very poor by comparison. The names are in a state of confusion too, with seed companies seeming to make up their own with gay abandon. Suffice it to say that only with

218

protection from the rain will they do well and many people keep them in the conservatory or greenhouse all summer and never let them out. The flowers are indeed quite extraordinary, being fully double, almost like coral. Whether this is in their favour is a matter of opinion, but if you should decide to give them a try sun and shelter are essential.

**Phacelia**

The sheer vibrancy of the blue in *P. campanularia* must be seen to be believed. The intense gentian shade is unique amongst summer annuals and so makes this plant indispensable to any but the most serious socialist who cannot disassociate the colour from its connotations! This, and its relative *P. tanacetifolia,* are American plants and really do prefer a soil that is not too rich, not too wet and in full sun. *P. campanularia* reaches little more than 9in (23cm) and can be given the usual hardy annual treatment, including autumn sowing — although thinning should be left till spring as there will inevitably be losses. The anthers are white and the leaves have a smell when bruised which is attractive to some people; it flowers all summer. *P. tanacetifolia* is much taller at 2–3ft (60–90cm), and has flowers in spikes which are a paler blue or sometimes slightly lavender in colour. This is not the most sturdy of plants and often needs a little discreet brushwood to keep it at its best. Both are splendid long-flowering plants for the annual border.

**Phlox**

*P. drummondii* is a much underrated little plant. It is an undemanding half-hardy annual, which responds well to sowing outside in May and quickly produces brilliant flowers, in some varieties with the most lovely bicolours. Perhaps it is the restricted colour range which does not inspire the enthusiasm other plants create or perhaps it is the slightly ungainly, stiff habit of the plants. However, with the improvement in varieties which has recently been seen perhaps they will become a little more popular.

They were introduced into Britain from Texas in the first half of the nineteenth century by, as you might guess, Drummond and the original plant was about 18in (45cm) high and deep rose in colour. Most of the current varieties are a great deal shorter at around 6–15in (15–38cm) and come in a wider range of colours. They thrive in sunshine and a rich but well-drained soil and a little acidity in the soil is no bad thing. They tend to give their best over a disappointingly short period if planted in soil which is too poor and dry. In the garden the taller ones are excellent long-lasting cut flowers while the smaller ones are ideal for edging, good in window boxes and tubs and fill pockets between early-flowering border perennials.

The varieties can be grouped into two — large- and small-flowered. Amongst the large-flowered, 'Beauty Mixed' is especially good growing to just 6–12in (15–30cm) and the flowers are around 1in (2.5cm) across. The colours tend to be dark and come in shades of red, purple, lilac, pink, yellow, scarlet, white and various intermediates. This is the only strain with any yellow, though it can tend to beige. 'Cecily' is altogether shorter and paler and many flowers are eyed. There are a number of different stocks of 'Beauty' about, some as tall as 12in (30cm), but if you want a taller version for cutting try 'Large Flowered Mixed', a more genuinely tall plant at 12–14in (30–38cm) in a similar array of colours.

In the smaller-flowered type 'Twinkles', at about 6in (15cm) has vast clouds of tiny star-like flowers in a good range of colours, many with a white picotee making 'twinkle' a very appropriate word. 'Petticoat' is an even smaller new introduction which although it starts to flower a little later than 'Twinkle' soon makes up for it in the brilliance of its display. The flowers are only ½in (12mm) across but the colour range is especially good and there are many bicolours. One or two plants of this variety, or 'Twinkle', could even be put in gaps on the rock garden without causing too much offence.

## Plumbago

An old established and well liked greenhouse shrub, *Plumbago capensis* is also planted outside in borders and containers. Recent seasons have revealed its worth. In the 1984 summer it did very well and even those raised from seed as well as cuttings-raised types provided a wonderful display of cool, pale sky blue, rather phlox-like flowers. In 1985, however, seed-raised plants hardly did a thing except make foliage, and that reluctantly, while cuttings-raised types proved adequate and no more. This is a rather twiggy and angular plant that even with staking rarely makes what you would call an elegant bush. It can reach 4ft (1.2m) in height but rarely does so in its first year. The flowers are rather delicate so do not put it with plants that are too bold. It can be flowered from seed in its first year if sown in January and given the usual half-hardy annual treatment but it helps if it is grown fairly warm and planted out as large as possible. It can also be raised from cuttings taken in late summer, overwintered in a frost-free greenhouse, grown on and planted out in late May or early June. If you want them to go up rather than straggle more or less sideways, then a stake is vital. It can also be trained as a standard if a seedling or tip cutting is allowed to run up to about 2ft (60cm) with any side shoots being kept short. It is then pinched and shoots in the upper third or quarter allowed to bush out. A stout support is vital.

There is a white form which is not available from seed and is not as appealing as the blue. Both make good, tolerant house plants too but outside like full sun, a fairly fertile soil and good drainage.

**Polygonum**

I encountered this plant, *Polygonum capitatum*, for the first time on the rock garden at Kew. It is a pretty little trailer making very flat growth of reddish foliage with small globular heads of tiny pink flowers for many weeks. It is a tender perennial which can be raised from seed in the first instance and then cuttings taken in late summer to overwinter for the following year. Its useful habit and restrained but very pretty flowering habits make it a good rock garden plant, good at the edges of paths and useful in baskets and tubs. As you might expect it likes sunshine and well-drained soil and is not happy in the shade, producing plenty of rather less red foliage and fewer flowers. Not a stunner, but worth a try and a good cold greenhouse pot plant for winter.

**Portulaca**

A group of succulent perennials usually grown as half-hardy annuals which are from parts of the Americas where they get a little more sunshine than is usual in Britain. Like gazanias they tend to open their flowers only in fine, sunny weather which does not make them universal favourites. There are two species grown, *P. grandiflora* and *P. oleracea*. The former has very narrow leaves and grows to 4–6in (10–15cm) and, although most varieties are of a noticeably trailing or spreading habit, there are one or two more compact types. 'Sundance' has semi-double flowers about 2in (5cm) across with nine colours in vibrant shades — the cerise is especially striking. Unlike many earlier strains 'Sundance' stays open in dull weather though sun and warmth are needed for a really stunning display. 'Cloudbeater' is similar though the flowers are a little smaller while 'Minilaca' is more compact, makes an excellent pot plant and is good as a close carpet in small containers. Light, well-drained soil is essential as is as much sunshine as can be arranged. Shade from buildings or trees will make growing them a waste of time and they are very poor in a bad summer. The usual half-hardy treatment is perfectly suitable but growth after germination can be slow and it is important not to overwater at that stage.

*P. oleracea* is even lower growing with rather broader, flatter leaves and slightly smaller single flowers. It is rather less fussy as to soil but still demands full sun and can be treated as a hardy annual, flowering in as little as eight weeks from sowing. 'Wildfire' is the variety to go for, indeed it may be the only one available, and it comes in a mixture of seven colours.

**Primula**

From this huge group there are just a few that are used in spring bedding and these are mainly polyanthus and primroses. But the situation is complicated by the plant breeders who, in recent years, have put most of their efforts into developing varieties suited to flowering in pots in cold or slightly heated greenhouses. Most of these are unsuitable for growing outside unless the soil conditions are almost as good as the compost they find in their pots when grown for Mother's Day.

There is only one widely available reliably hardy polyanthus, 'Crescendo'. Although the 'Pacific Giants' are often grown, having been bred in a climate a good deal warmer than in Britain, they are not reliably hardy in Britain although in containers which are given protection from the worst of the elements they will often thrive. It is unfortunate that it is the 'Pacific Giants' which are usually to be had in a range of single colours whereas only the blue from the 'Crescendo' colours is found. Especially hardy types were once increased by division but these are now only rarely seen. In milder areas the 'Pacific Giants' will do well as will a number of other strains.

Amongst the primroses the choice is as limited. There is not one that has been bred to thrive outside, but 'Mirella' seems to be the best of those bred for pot growing at coping with outside conditions. 'Spring Charm' is good too, as well as the newer 'Wanda' types which are very small but tough with good-sized flowers on short stems. Look out for the 'Wanda Hybrids' and 'Panda'.

To encourage any of these varieties to give their best do not grow them in cold exposed places. Improve the soil to create a well-drained growing medium which nevertheless will not dry out. Small areas can be opened up with peat and sharp sand or even perlite. Alternatively, beds can be made up with a mixture containing three parts used John Innes compost, one part sharp sand and one part moist peat. Avoid choosing limey sand and always use moss peat to avoid inadvertently increasing the lime content. If the soil conditions are right and reasonable protection given against frost and cold winds you will have the best chance of success. Tubs and troughs offer perhaps the most reliable method of cultivation as the compost can be chosen to suit the plants. A mixture of commercial ericaceous and standard mixes is ideal or you can make your own using the Chempak Ericaceous Pack rather than the standard potting mix in your mixture.

Germination of primroses and polyanthus is a little more tricky than that of most plants treated as biennials and with the seed being fairly expensive the protection of a greenhouse is useful. A little warmth will encourage quick germination which will ensure the maximum total germination. Use perfectly clean pots or trays and a peat-based seed compost; avoid John Innes Seed Compost for primulas. Sow thinly and leave the seed uncovered. If you feel that you will be unable to keep the temperature constant or make sure the compost surface is always moist, give a very light covering. A temperature of 60–65°F (15–18°C) is necessary; a higher temperature will reduce germination while a lower one may also result in an overall low germination rate. It is vital that the surface of the compost does not dry out for even an hour. Cover the pots with white polythene or glass and paper and as soon as the root starts to emerge from the seeds sift a little compost over the surface. When the seedlings emerge remove the cover and avoid direct sunlight or drought. Prick out the seedlings into pots and, when established, the pots can go outside in a cold frame until autumn planting time.

Use polyanthus in pots with dwarf tulips, dwarf daffodils or hyacinths and primroses with smaller bulbs. In the garden, polyanthus make good edgers while primroses are often best in mixtures in small beds of well-prepared soils.

**Pyrethrum (Silver Feather)**

The flowering pyrethrums unfortunately will not usually bloom in their first year from seed so the only member of the genus which fits in here is *P. ptarmicaeflorum*, the silver feather, grown for its almost white foliage. It is an upright plant growing to 12–15in (30–38cm) with flat fans of very finely dissected foliage in greyish white. Although a

*Figure 7.17* Pyrethrum ptarmicaeflorum

perennial it is treated as a half-hardy annual and the usual half-hardy treatment serves it well. In the garden it is happy in most soils and situations although it tends to get rather leggy in shady spots. It is a particularly useful plant in tubs where its upright habit is most suitable and for a lovely combination try it with *Salvia farinacea* 'Victoria' and 'Rose Picotee' or 'Resisto Rose' petunias. With white flowers it is also very pretty — try *Antirrhinum* 'White Wonder' and *Petunia* 'Recoverer White'. Another good companion — with *Salvia* 'Laser Purple'. For a combination solely with foliage grow it alongside 'Bravour' parsley and *Hibiscus* 'Coppertone'.

# R

**Reseda (Mignonette)**

No one would suggest that these are the most ornamental plants in the world — but the scent! Of course it is for the lovely sweet scent that they are grown, and as the plants are so easy they should be in more gardens. It was once common in cottage gardens and was also grown especially for cutting and even sold in florists. The name comes from the French 'mignonette d'Egypte' meaning 'little darling of Egypt' which relates to its wild home and the fact that Napoleon is said to have sent seeds to Josephine, who was a keen gardener. It is a hardy annual for sowing in spring or autumn and can also be grown in the greenhouse in winter for cutting for the house. It grows to about 12in (30cm) in a slightly open manner and the flowers are generally a rather greenish colour although there have been some improvements. 'Sweetly Scented' is slightly rusty, 'Red Monarch' has red and green flowers, 'Goliath' is brilliant red. As it is the scent that is really important I always go for the 'Sweetly Scented' or 'Fragrant Beauty'. Last year I grew 'Giant' and 'Sweetly Scented' and they were exactly the same — except that 'Giant' smelt more strongly! It looks surprisingly good in a pot with *Lobelia* 'Red Cascade' and is worth popping into window boxes and tubs to give a waft of scent to otherwise unscented plants. A slightly alkaline soil is preferred, in sun or part shade.

**Ricinus (Castor Oil Plant)**

When visitors first come through the gate they usually shriek: 'What on earth's that?' This query can refer to a number of plants but it is usually the onopordon or the ricinus to which they are referring. Particularly as it is usually the purple-leaved version of the castor oil plant that I grow. But first let me clear up a confusion that has arisen about the name. The true castor oil plant, from which the oil is actually derived, is indeed *Ricinus communis*. The plant to which the name is also sometimes given is × *Fatshedera lizei*, a cross between *Fatsia japonica*, a large more or less hardy shrub with enormous dark, divided leaves and the Irish ivy, *Hedera helix* 'Hibernica'. × *Fatshedera* is often grown as a pot plant, especially in its form with irregular yellow margins to the leaves. The leaves do resemble those of the true castor oil plant but they are much smaller and thicker.

Castor oil is extracted from the very attractive, large, speckled seeds. To raise plants for bedding, the pretty seeds are sown individually in peat pots at a temperature of not less than 70°F (21°C) in March and grown on steadily. If grown warm they can be moved into 5in (7.5cm) pots

*Figure 7.18 The roots of ricinus soon grow through the peat pots in which the plants were raised*

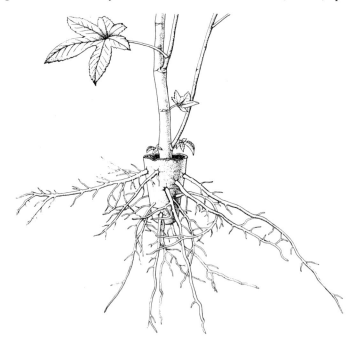

before planting out at the end of May. They will probably need staking at this stage.

They are best grown in full sunshine and a slightly gravelly or sandy soil does not seem to bother them although they are more likely to reach a good size, say 4–5ft (1.2–1.5m), if the soil is good. This plant, more than most, benefits from a liquid feed as the impact really does depend on their reaching a good size and making strong, well-branched specimens.

The best variety for coloured foliage is undoubtedly 'Impala' which produces leaves of a shining reddish purple as well as especially showy clusters of yellow flowers and spiky red seed clusters. There is also a strain usually grown under the name of 'Zanzibariensis' which contains not only this reddish purple shade, but also bronze and plain green. The flowers are rather less dramatic. A more recent introduction is 'Mizuma' which has green leaves but red stems and red veins in the leaves.

But it is 'Impala' which really sets the standard these days and it makes happy combinations with a number of other plants. Go for dramatic contrasts or hot, fiery blends. Plant one at the foot of an overwintered *Eucalyptus globulus*, if you have such a thing, with some big white tobacco plants (*Nicotiana alata*) at the side and maybe some of the larger silver foliage plants like *Cineraria maritima* 'Diamond' plus even some white petunias. Alternatively go for the reds, yellows and hot oranges. Red dahlias like 'Coltness Scarlet' or orange like the vegetatively-raised 'David Howard'; try *Eccremocarpus scaber* trained through the branches so that orange trumpets peep through, *Lobelia cardinalis, Tithonia* 'Torch' and 'Yellow Torch', almost as big, coppery marigolds like 'Scarlet Sophie' or 'Cinnabar', red salad bowl lettuce, *Pyrethrum* 'Golden Moss' — I am sure you get the idea. It is actually a perennial shrub in warmer climes but I dare you to overwinter one in the greenhouse and see what happens the following year; it will probably not let you in the door.

## Rudbeckia

Rudbeckias are very much a British speciality with the three best varieties all bred in Britain by Ralph Gould at Hurst. They are hardy plants which are treated as half-hardy annuals and can be set out a little earlier than the true half-hardies, if properly hardened off, to give a little more space in the greenhouse and frame. The flowers are big, gold and daisy-like and usually in various shades of orange; one at the top of each stout, rough stem. They are amongst the best of seed-raised cut flowers, lasting very well in water and some of the darker bronze and mahogany shades are especially useful in arrangements.

The 'Rustic Dwarfs' grow to about 2ft (60cm) and come in a lovely range of colours from red, bronze and mahogany through to gold, yellow, orange and with some bicolours. They all have a contrasting black central cone and flower all summer. 'Marmalade' is pure golden yellow with a large dark centre, the flowers sometimes being as much as 5in (12.5cm) across. A real dazzler which is a wonderful bedder giving a marigold colour but in a dramatically different style; it is a good cut flower too. Finally, the most recent Hurst variety 'Goldilocks'. The flowers are rich yellow, about 3in (7.5cm) across, and come in a range of double and semi-double forms, of which the doubles are especially impressive. This variety won a Fleuroselect Bronze Medal in 1985. Its other attributes are that it flowers early on quite small plants, keeps going merrily all summer, is upright and neat, and is a splendid small garden plant.

Strictly speaking these plants are biennials and if you need an especially early and prolific display, for cutting for example, you can sow in a cold frame in summer and plant out in autumn. They are more elegant than taller African marigolds, with more interesting forms and fit far more comfortably into the home garden. 'Goldilocks' is good with purple hibiscus, *Salvia* 'Victoria' or the more fiery tuberous begonias.

Rudbeckias are best sown in about February and after pricking out the temperature can be reduced to 50°F (10°C), or even slightly less, for growing on, and they will still make good plants and flower early.

# S

**Salpiglossis**

No one in their right mind would suggest that salpiglossis is an ideal bedding plant. It is sad but true that in Britain we just do not have the climate for it to do very well outside. Salpiglossis is related to petunia and nicotiana but is the least suitable of the three for outside. Present varieties make rather spindly upright plants growing to about 18in (45cm) but even on these it is obvious that the flowers are rather special. More or less like a petunia in shape, though a little smaller, they come in astonishing shades of red, yellow, bronze, violet, purple and blue with dramatically contrasting veins. Unfortunately in the British climate they tend to blow over, there is insufficient leaf to make an elegant plant, the flower production is poor. Those flowers that are produced tend to be very easily battered by rain and torn by wind. They make ideal cold greenhouse pot plants and in this situation the weather cannot harm them. But in spite of the enthusiastic eulogies of optimistic seedsmen who need to up the profits, only in sheltered gardens in favoured parts of the country are they worth trying. If you insist on having a go try 'Diablo', or 'Ingrid' — both mixed strains. Remember, though, that although they are susceptible to bad weather they do not like to be baked and neither do they like their roots disturbed too much so prick out into small pots.

But better things are on the way. At least one flower breeding company in Britain, and others elsewhere, are doing their best to improve them. The company that produced 'Domino' nicotianas and 'Calypso', 'Viva' and 'Malibu' mimulus is working on improving them sufficiently to make them generally useful as garden plants.

## Salvia

The brilliant scarlet bedding salvia has long been a favourite of parks departments but also of home gardeners, who use it to create a dramatic blast of colour to blind motorists and cause frail old ladies to reach for their smelling salts. The old red, white and blue with alyssum and lobelia still makes an effective bed of colour although fashions change and it is not to everyone's taste these days. These familiar scarlet-flowered types are derived from *Salvia splendens*, a perennial shrub from Brazil which in the wild grows up to 3ft (90cm) high, although now invariably treated as a half-hardy annual and rarely reaching more than 15in (38cm), and often much less.

The varieties have developed into two types, one group has pale foliage with rather stubby spikes set low on the foliage and often at an angle. The other group, which is becoming more common, has much darker foliage with taller spikes carried in a more upright manner. For many years 'Blaze of Fire' was the standard variety. It is still around and at least you get plenty of seed in the packet for a reasonable price. There are a number of strains of it now, and you will not know which particular one you are getting when you buy your packet, but it will be a tough, pale-leaved type, probably without the impact of the best of the modern dark-leaved varieties. 'Red Riches' is the one to judge them all by now. It is a French variety with the most vivid of flowers in large numbers over dark green foliage. You may also find it listed as 'Ryco'. Others to look out for which are also good are 'Firecracker' ('Fusco') with especially dark red flowers and 'Meriyanne' which is not available at the time of writing; but with more flowers than any other variety I have seen it cannot be long before it appears. There is just one colour other than red which seems to work with salvias and that is purple — and just one variety, 'Laser Purple' stands out. It is a lovely rich, deep shade and very long-flowering. Various companies have introduced paler varieties and the most commonly seen is the hideous 'Dress Parade', a mixture of sickly pinks and salmons, cream and watery purples which is enough to put anyone off gardening for life. There is still work going on in this group and I must say I have seen quite a good yellow which may appear in a year or two.

Another species has produced a couple of varieties in rather a different style, and although they will never have the universal appeal of the *S. splendens* group, they are nevertheless very useful in the garden. They are derived from *S. farinacea*, a rather less coarse, upright species from Texas. At present one variety is fairly widely seen, following its Fleuroselect medal in 1978. This is 'Victoria', which reaches about 16in (41cm), and has grey green foliage in a neat bun about half the height of the plant and tall narrow spikes of dusky deep blue flowers on dark purplish stems. This is unlike any other summer plant and deserves to be much more widely grown. It is lovely in the garden with silver foliage, pastel busy lizzies or soft yellow marigolds like 'Solar Sulphur'. There are two white counterparts called 'White Porcelain' and 'Argent' which are also worth growing and a new mixture of dark and pale blue plus white, but these are rarely found. All these varieties produce small tubers during the summer and in theory you could lift them in the autumn and store them like dahlias; try it if you like, but they are not difficult to raise from seed so there is little point.

The next in popularity amongst the half hardies is *S. patens*. A Mexican plant, reaching 2ft (60cm), the flowers are large and the most stunning rich blue but spaced out rather loosely on the stem. It is more of a candidate for the mixed border than the bedding scheme, the display being rather open in character. In the 1950s, there was a lovely pale variety, 'Cambridge Blue', grown, and this has recently been resurrected and should be available again in the next year or two. This makes a lovely swirling sea of soft blue colour if grown in a large group.

There is also a hardy annual salvia which is rather different from the half hardies in that it is not grown for its flowers, but for its leafy bracts which surround the flowers. *S. horminum,* sometimes mistakenly known as clary, is a very upright plant growing to about 18in (45cm) and whose actual flowers are small and insignificant. But underneath each tiny cluster of flowers on the stem, are two leafy bracts which may be blue, pink or greenish white, each with much darker veins. It makes a soft and interesting garden display and is also dried for indoor winter arrangements. Most of the mixtures have names like 'Art Shades' and 'Bouquet Mixed' but a new rather shorter strain has recently appeared called 'Claryssa' which grows to just 15in (38cm) and is better branched and with more intense colours than older types. It comes in separate colours too. Another mixed border or annual border plant, not a bedder, and lovely with pale grasses in winter arrangements in the house.

The true clary, *S. sclarea,* is a biennial or short-lived perennial making big silvery rosettes in its first season followed by huge clouds of blue, or slightly pink, and white flowers. It can be grown as an annual if sown early but you will have less of those broad, felted rosettes before the flowers appear.

Growing salvias from seed is not as straightforward as you might think. You can sow in mid-February if you want to pick them out into pots for plants, in flower, to put in containers, or leave it until mid-April if you want a lot of plants to put out without flowers from trays. For germination to be good the seed must be constantly moist and so to avoid damping off a very well-drained seed compost is needed. The temperature should be constant too, 65°F (18°C) is warm enough but if it varies you will find that seedling emergence is erratic. After pricking out, the temperature can be reduced to about 55°F (12°C) but if it falls any lower, especially early on, then plants may be stunted and foliage may look sickly. Care when pricking out is important and steady hardening off rather than a sharp shock also pays.

Clary is best sown in pots and pricked off early into 3in (7.5cm) pots to be planted out as soon as the roots start to fill the pot. If you leave a few seed heads on the plants seedlings will appear in the border and as long as they are moved when still small, any in inconvenient spots can be found more appropriate homes.

**Sanvitalia**

A widely spreading and low growing little hardy annual, this is a close relative of the zinnias. It only grows to 6in (15cm) but can spread out to 18in (45cm) although some breeding work has been done to try and reduce the spread — another tendency to marigoldisation; these plants have not yet appeared on the market, thank goodness. The flowers are bright yellow each with a black centre and about ¾in (2cm) across. They like full sun and are happy in any soil which is not too wet. The usual hardy annual treatment suits them well and they can be autumn sown in mild areas although it pays to leave the thinning until spring. Grow them as edging to broad paths, on corners of annual borders and in gaps in beds. They also make good light ground cover plants, ideal for surrounding marigolds or dwarf calendulas. They are a little gaudy for the rock garden which is often suggested as being a good home but they could be worth trying in baskets and tubs where they would probably trail down prettily. There was once a double-flowered form which has vanished from cultivation, although an orange may appear before long.

**Scabiosa
(Sweet
Scabious)**

Splendid hardy annual cut flowers growing to over 2ft (60cm) carrying mainly double flowers in reds, pinks, lilacs and white together with an interesting very dark maroon. The names are not very inspired and I am afraid this may well reflect the lack of interest in these plants. 'Double Mixed' grows to about 2½ft (75cm) while 'Dwarf Double Mixed' makes just 1½ft (45cm). The usual hardy annual treatment serves them very well, but germination can take over a month so be patient. Only in mild areas does the sweet scabious overwinter happily from an autumn sowing. Give them a light, well-drained soil in full sun and they will produce flowers for cutting, which last very well in water, all summer. Scabious also makes a soft, restrained contribution to mixed borders and groups of annual plants.

There is also a species grown for its seed heads which are dried for winter decoration. This is *Scabiosa stellata*, sometimes given the variety name 'Drumstick', 'Silver Moon' and 'Paper Moon'. This too is a hardy annual growing to about 1½ft (45cm) whose pale blue flowers are followed by 3in (7.5cm) rounded heads of cup-shaped papery bracts in a pretty beige shade. The heads can be left to dry on the plants and need no special treatment.

Do not let that word in brackets at the top put you off, no one is going to suggest you put *that* in your garden. Rather, a slightly more delicate annual from South Africa (again) with strong, purple ray petals and a yellow centre and as easy to grow as you might expect. It grows to 1½–2ft (45–60cm) and flowers all summer especially if you can be bothered to take the dead flowers off. Although always said to be a half-hardy annual it is not really the sort of plant to raise in trays and plant out 9in (23cm) apart. Better to sow in late April where it is to flower, thin out to no more than 3in (7.5cm) and let it make a tall yellow and purple mound. It will sometimes seed itself if you are less than assiduous with the dead-heading. Most soils suit it well, although sun is preferable. In wet seasons the plants get very leafy.

**Senecio (Ragwort)**

A brilliant orange relation of the poppies growing to 1½–2ft (45–60cm). As you might expect, it likes a well-drained soil but is surprisingly happy in fairly shady spots. It is best sown in March or April where it is to flower as it resents being moved. The leaves are also very fragile so the minimum of disturbance is helpful. The flowers have four big petals each with a maroon blotch at the base and a cluster of yellow stamens in the centre — very bright and cheerful.

**Stylomecon (Wind Poppy)**

# T

Marigolds must be the most reliable of all bedding plants. Anyone can grow them — the seeds are big, they germinate quickly, they flower quickly and whatever the variety, the display will be bright and long. They are traditionally divided into African (*T. erecta*) and French (*T. patula*) types — although both are derived from wild plants that grow in Mexico! Africans tend to be tall and bushy with large numbers of double flowers while the French ones tend to be shorter and more spreading with smaller single, semi-double or double flowers in rather larger quantities. In many ways the two groups have become less distinct over the years and now the merging is complete with the introduction of the best group of all for the home gardener, the Afro-French marigolds. These varieties combine the qualities of both with the addition of an extra bonus.

**Tagetes (Marigolds)**

The plants are low growing but vigorously spreading and they flower very early, some in as little as five weeks from sowing. And once they have started to flower they should persist right through to the frosts. The only problem with such enthusiastic blooming is that sometimes they flower themselves to an early death before summer is over. There are two main groups of Afro-French marigolds making the running now. One originates in Britain and the other in France. The British are mainly single-flowered, the French mainly double.

The first of the single-flowered types, 'Nell Gwynn', was bred and introduced by Asmer Seeds of Leicester in 1980 and in five more years of breeding work, another four singles as well as three doubles have been introduced. The doubles are much less easily available to the home gardener.

The way they are produced is interesting. The seed in the packet is the result of crossing an African marigold with a French marigold. Only the African parent is used to produce seed and in addition to its other qualities it is specially chosen to produce no pollen so that self-pollination is impossible. The French parent provides the pollen for fertilisation but no seed is collected from it. When the seed is collected the variety that has been created should be very uniform, whatever the parents are like. The parents used are developments of old or existing varieties which the plant breeder, through a mixture of experience, calculation, intuition and inspiration suspects will produce the desired result. Many crosses between likely plants are tried on a small scale and those which produce promising offspring repeated on a larger scale and varied slightly until the ideal variety is created.

One very valuable side effect of this particular hybridisation — and this is where the bonus comes in — is that because of their complicated genetic make-up this particular group of varieties never produces any seed. The result is that plants flower far more prolifically, as if they were trying harder and harder to produce the seed which never actually materialises.

There are, though, a couple of minor defects. Occasionally, totally abnormal plants appear, usually only at the rate of one or two in a hundred but they do spoil the display. They are noticeably taller and more upright, with very dark, dense foliage and they do not usually flower before the autumn. Afro-French marigolds should flower early and all the plants of each variety should start flowering together, usually before planting. So if there are one or two that have no flowers or flower buds when you put them out, it is probably safest to discard them. The other problem is that germination is not always as good as with other marigolds although this is not now such a problem as it once was.

The varieties in this single-flowered group are:

'Nell Gwynn' — Yellow with a deep red central splash.
'Suzie Wong' — Bright golden yellow.
'Mata Hari' — Gold with a red throat.
'Josephine' — Brilliant orange.
'Little Nell' — Dwarf version of 'Nell Gwynn'.

There are also three double-flowered varieties: 'Beau Brummel', a rich gold, 'Beau Geste', orange with a red splash and 'Beau Nash' in dark red.

Just four French varieties are available so far although more may soon appear. They are all fully double-flowered, very resistant to bad weather and the flowers are up to 3in (7.5cm) across. All have the prefix 'Solar'.

'Solar Gold' — Rich golden yellow crested flowers.
'Solar Sulphur' — Pure sulphur yellow.
'Solar Orange' — Orange with a slight red flecking.
'Solar Topaz' — Bright golden yellow, paler than 'Solar Gold' and crested.

The doubles seem to have captured the gardener's eye more than the singles although the singles are especially spreading and make excellent weed supressors.

Two older Afro-French types are also worth considering. 'Sunrise' is a golden double with a red throat to the petals and 'Red Seven Star' (sometimes listed as 'Seven Star Red'!) is an intensely double mahogany variety whose 3in (7.5cm) flowers almost hide the foliage. The first flowers of some of these earlier types may not be typical of the flowers that appear later.

Although not the cheapest of marigolds these groups will make a dramatic impact in the garden. Their persistent flowering makes them especially valuable in smaller gardens and in containers where every flower counts, and they are sufficiently tough to be raised from seed sown in a cold greenhouse in April and still be flowering by the time they are planted out.

With such a large group, comprising so many varieties with more introduced each year, it is wise to look at them systematically. I must say that the vast majority of marigolds on the market, be they French, African or various intermediate types, will produce an exceptional garden display and it may well profit the gardener were the plant breeding companies to devote some of the vast resources that go into producing yet another new marigold to developing other plants. And the tendency to develop other plants like tithonias so that they look like marigolds is even more wasteful. Still, at present we can only grow what exists — so here goes.

*Dwarf French Carnation Flowered*

These grow up to 12in (30cm) high and have uncrested double flowers. Few resemble carnations in the slightest but this seems to be the term used in catalogues. The 'Queen' series, also known as the 'Sophia' series (just to confuse you) are probably the best garden plants in this group. The flowers, sometimes described as 'Camellia flowered', are 2in (5cm) across and the petals especially broad and neatly overlapping. There are four colours in the mixture which are sometimes found under their own names. 'Queen Sophia' has flowers which open bright

*Figure 7.19* Tagetes *'Red Cherry'*

coppery red with the finest orange edge and then develop a more orange shade as they age. Bad weather damages them hardly at all. 'Queen Beatrix' has bright yellow petals with a red throat, 'Scarlet Sophia' starts off in a russet shade with a fine orange edge and develops a more fiery red tint as it ages — the reverse of 'Queen Sophia'. A new one which has yet to reach the catalogues is 'Honey Sophia', which is a russet colour edged with a rather broader, yellow edge that changes shades, yellowing steadily. Others in the group worth looking out for are the 'Bonita' varieties reaching only 8in (20cm), and 'Red Cherry' in russett with a fine yellow edge. All are good at tolerating bad weather well.

*Dwarf French Crested*

Single or double flowers with the central area raised in a crest of tightly rolled petals. You either like them or you don't and many do. The petals and crest are sometimes of contrasting colours, especially in the 'Eyes' series. A single row of reflexed petals surrounds a large dense crest making a dramatic flower. The plants are about 12in (30cm) high and there are three versions. 'Tiger Eyes' is the most striking, the petals in scarlet with a crest of orange yellow. 'Lion Eyes' has orange petals with a red basal blotch and an orange crest while the curiously named 'Familiar Eyes' has clear yellow petals with a small red basal blotch and a yellow crest. Very striking, but the first variety, 'Tiger Eyes', is outstanding.

The 'Boy' series is well established as the top double crested variety and most companies list some or all the colours. It is the shortest, reaching only 6in (15cm), and produces a constant succession of flowers. There are five colours known, in the way of these things, as 'Sunny Boy' — gold, 'Orange Boy' — deep orange, 'Yellow Boy' — yellow, 'Spry Boy' — yellow crest with mahogany collar and 'Harmony Boy' — deep orange crest with mahogany collar. The mixture is available everywhere and is known by the rather obvious but curiously spelt name of 'Boy O'Boy'. The orange was a Fleuroselect Bronze Medal winner in 1979. Others in this crested group to look out for include 'Yellow Jacket' with larger flowers than 'Yellow Boy' and more of them, also a Fleuroselect Bronze Medal winner, in 1982; it has very tight, brilliant pure yellow flowers on 6in (15cm) plants. There is also 'Honeycomb' at 10in (25cm), yet another Fleuroselect Bronze Medal winner, this time in 1975. The flowers are 2½in (6cm)

237

across and mahogany broadly edged in gold. Very dramatic.

*Dwarf Single French*

The two Mariettas are old favourites in this section. They are amongst the very few now listed in catalogues which are not F1 hybrids and so are also amongst the cheapest. 'Dainty Marietta' grows to about 7in (18cm) on neat plants and the flowers are yellow with a crimson central blotch. 'Naughty Marietta' is rather taller with a red central blotch. 'Pascal' is similar to 'Naughty Marietta' but only 6in (15cm) high and the newer 'Leopard' is more yellow and spreading. But there are two which are outstanding in this group, 'Cinnabar' and 'Silvia'. 'Cinnabar' has very large flowers in about as red a shade as any marigold on 12in (30cm) plants that are bushy and covered in flowers. Simple and very useful. 'Silvia' was a Fleuroselect Bronze Medal winner in 1982 and has large, golden flowers with a distinct wave to them. Very impressive and does well in the shade.

*Dwarf African*

In recent years African, or American marigolds as they are sometimes known, have been getting shorter and the flowers have been getting bigger in proportion. The problem has been that the big, fully double flowers have often suffered badly in wet summers with botrytis rotting the centres of the dense flower heads. Two groups catch the eye in most seasons, the 'Ladies' and the 'Incas'. The 'Lady' series has been around for a while and is a firm favourite. The plants are 12–15in (30–38cm) high and covered with densely double flowers. There are four varieties — orange, yellow, gold and primrose and a mixture of all four called 'Gay Ladies'. The only problem is that like a lot of these dwarf African types the plants are rather upright and blocky but the power of the colours is tremendous.

A more recent introduction is the 'Inca' series — gold, orange and yellow with especially dark foliage. They reach just 12in (30cm) and the flowers can be as much as 4in (10cm) across. There have been some dramatic differences in their tolerance of bad weather and some steadfast disagreement amongst experts as to their value, but there can be no doubt that in a good summer you need sun glasses to keep out the glare — and not from the sun. In my experience the heads are so dense that rain does not

238

penetrate. The buds tend to stay below the flowers before opening so they do not spoil the display.

*Tall African*

Fewer varieties over 18in (45cm) are listed now as the breeders have been reducing the height, but in some situations really tall plants are just what you need. At 2ft (60cm) there is the 'Jubilee' series — gold, orange and yellow — which although tall are still rather dense in growth. The flowers are bigger than those of the 'Incas' and the plants are more open. Going up a size further, there is 'Doubloon' at 3ft (90cm) with pale yellow flowers up to 5in (18cm) across; huge they are but you would hardly call them elegant.

*Hybrida*

There is only one variety in this section, 'Florence', which is a cross between a 'tagetes' (*T. tenuifolia*) and an African marigold — you may find it listed under either. It makes a plant about 15in (38cm) high with rather upright growth and bright orange semi-double flowers. It is a Fleuroselect Bronze Medal winner and an interesting and useful addition to the marigold family.

So, having got this far and simplified the vast range of varieties, a few words about raising them. They are half-hardy annuals which can be raised from a relatively late sowing (never sow before March), and which germinate and grow quickly.

Most of the smaller varieties and especially the vigorous triploids can be sown in a cold greenhouse in April and still flower very soon after they are planted out. The modern African types can be very slow to establish if planted out in flower so this again is a reason to sow late. The triploids are also worth sowing outside in June to fill any gaps in mid-summer where unforeseen accidents have taken place; thinned promptly they can be watered, dug up and replanted in the border where they will soon grow away and flower merrily as long as they do not dry out.

In the garden they most assuredly prefer to be in the sun. In fact they are so brilliant in colour that even though they are far from their best in the shade they will still make a reasonable show. The soil should be fairly fertile and, although the plants will stand some drought, they do benefit from a drink in dry spells. All except the tallest Afro-horrors, and they can be a bit gross with their vast garish flowers, are excellent container plants because of their long-flowering qualities and the brilliance of their display from such a small space.

Making suggestions as to combinations with other plants could fill up the rest of the book as although the colour range is restricted to mahogany, through a sort of scarlet to orange, gold and primrose, each tone demands separate treatment. The range was amply demonstrated by the dramatic Sunburst display at the Liverpool Garden Festival.

This was a triumph of preparation and co-operation by the British bedding plant industry in which 40,000 marigolds were planted out in beds totalling well over 2,000 square metres in area. And it was not just planned on paper. The previous year a trial was done to find the varieties that would be most suiable and plants were grown both on the site of the proposed display and at a Ministry of Agriculture research station near Derby. Not all the varieties are available to the home gardener but those that came out best in both situations, where conditions were noticeably different, were 'Little Nell', 'Moll Flanders', 'Suzie Wong' and 'Beau Geste', 'Solar Orange' and 'Solar Gold', 'Sunrise' and 'Honey Bee'. Quite a success for David Haswell, plant breeder at Asmer Seeds, who not only bred Asmer's single and double triploids, but while with Suttons before joining Asmer, also bred 'Sunrise'.

The plants gardeners call tagetes (*T. tenuifolia*), to distinguish them from marigolds — they are all tagetes to the botanist — are altogether more dainty in appearance with fine, feathery foliage and small single flowers which appear in great numbers. The plants tend to make almost round balls of colour. The colours are in the lemon to rust range with 'Golden Gem' at 6in (15cm), 'Tangerine Gem' at 9in (23cm) and 'Paprika' at 6in (15cm), all having self-explanatory names. 'Starfire' is a mixture of colours with the disadvantage of being rather uneven in height. All varieties make excellent edgers with the great advantage of creating a very definite solid barrier between the rest of the border and the grass.

**Thunbergia (Black Eyed Susan)**

Tender climbing perennials almost always grown as half-hardy annuals, thunbergias are slender climbers for sheltered spots outside and for the cold greenhouse or conservatory. They reach about 4ft (1.2m) with small heart-shaped leaves and very showy flowers carried singly up the stems. In the variety from which the plant gets its common name these flowers are orange with a dark purplish black centre but in the most widely available mixture, 'Susie', the flowers may be pure orange, pure yellow, pure white, orange with a yellow eye, and yellow or orange with a dark eye. There is also a much larger-flowered variety, 'Angel Wings', with blooms up to 2in (5cm) across and pure white flowers with a yellow eye. Treat them all as half-hardy annuals, but sow in small peat

*Figure 7.20*
Thunbergia alata

pots and transplant to 3in (7.5cm) pots as they do not like being disturbed. Three plants can go in a 7–8in (12–20cm) pot, one in a 5in (12.5cm) pot, for the greenhouse or conservatory.

Outside they prefer a relatively sheltered spot with some fine mesh or trellis for support. They make excellent container plants and a plant or two in the back of the tub in a sunny porch, given a little support, will produce a lovely display in the sun, shelter and good compost that they prefer. In hanging baskets too they can be very successful, given that they are not set in the sort of windswept spot that baskets are often sited. But do not attempt to get the plant to trail; rather set a plant at the base of each chain and train them up the chain and on to the bracket so disguising the essential ironmongery. It is worth remembering that thunbergia is a plant especially prone to damage from red spider mite, both in the greenhouse during the early stages and later outside, as the sort of warm sheltered corner the plants prefer also provides a happy home for the mites.

**Tithonia**

The first time I ever grew tithonia, on extremely fertile Fen soil — you should have seen the crop of nettles which was there previously — I was astonished. Only one of the five plants I put out survived owing to the unexpected predations of slugs but this one eventually made a tall bushy plant 5ft (1.5m) high covered in big single orange flowers; very impressive it was too. This variety was called 'Torch' and there is now a bright yellow version called, as you might guess, 'Yellow Torch'. Unfortunately, plant breeders have started to try and marigoldise it (it is a close relation) by reducing the size somewhat and the latest variety 'Goldfinger' (also known as 'Torch Improved'!), only reaches the relatively modest height of 2½ft (75cm). It is far less impressive. Of course with its long rather hairy, undivided leaves it will never look quite like a marigold but the intention is one not to be encouraged. The seed is much like that of the marigold and germination is reliable but prick them out into 3in (7.5cm) pots as they are not slow to get going and spaced out a little on the greenhouse bench will bush out more effectively. They like plenty of sun, and will not do well in quite such poor soils as marigolds. In exposed gardens a stake may be a wise precaution.

**Tripteris**

Although technically this is probably an osteospermum (or even a dimorphotheca) the catalogues list it as tripteris (so named because of its three winged seeds) and so it appears here. It is a brilliantly coloured member of the daisy family with intensely orange flowers around 2in

242

(5cm) across each with a dark blue-black eye. The plants are about 18in (45cm) high and once planted out flower all summer. They sometimes close in dull weather but there is ample compensation in the brilliance of the display in sunshine. They last well in water, have a fresh hay-like fragrance in the foliage, and this is altogether a much underrated plant. It is not entirely hardy so is best given the usual half-hardy treatment, although it can be sown outside in May. During dull days the ray florets roll under towards the centre making the flowers look dead; they then unroll again in sunshine. So dead-head in sunny weather when it is far easier to distinguish the dead flowers from those still fresh. The variety 'Gaiety' is the one usually found.

## Tropaeolum (Nasturtium)

Always reliable, nasturtiums are amongst the easiest and cheeriest of plants and there is more to them than simply their appearance. The nasturtiums we grow today are derived from three South American species which over the years have been crossed with one another, rigorously selected and refined. Perhaps the most remarkable change is that varieties making rounded mounds of foliage 9in (23cm) high have been developed from a plant which in its original form can grow to 15ft (4.6m).

There are just two varieties around which are close to the original wild type, *T. majus* and *T. lobbianum*. The 'Tall Mixed Hybrids' are just that. The flowers come in various yellowy and reddish oranges, not a great range, and they are extremely vigorous. I have had a plant go to the top of a 12ft (3.6m) laburnum tree, run across the top and then fall into the hedge nearby and set off along it. The other climbing one is a variety of *T. lobbianum* called 'Spitfire'. This Colombian species has more tubular flowers than other nasturtiums and the variety 'Spitfire', the last remaining of a series of hybrids, has reddish stems and orange flowers with red markings in the throat. The leaves are smaller too and the leaf stems shorter so the flowers stand out well.

Coming down in size dramatically there are the 'Gleams'. Described as semi-trailing rather than bushy, they grow to about 15in (38cm) high with short trailing shoots making them excellent ground coverers and also useful in window boxes and baskets. Their semi-double flowers come in various oranges, yellows and reds plus salmon, cerise, primrose. This is a variety where soil and growing conditions exert a dramatic influence. As well as the mixture there are orange, gold and scarlet colours available separately.

It is in the dwarf types where there has been the most change, with the aim of turning nasturtiums into bedding plants. Generally speaking, this has been successful but does not seem to have filtered through to gardeners, far more could grow and enjoy them. The state-of-the-art in dwarf nasturtiums is probably 'Whirlybird'. This variety reaches just 10in (25cm) and has two unusual characteristics — the flowers are without the normal spur and they also face almost directly upwards instead of sideways, which on a plant growing to less than 1ft (30cm) dramatically adds to the intensity of the display. The flowers stand well above the foliage too (this foliage is also smaller than average) and comes in various sparkling oranges, yellows and reds. The intense scarlet is often available separately.

'Alaska' is another good dwarf type and this too has a unique characteristic, its white speckled foliage. This of course helps restrict its vigour and the flowers generally stand above the foliage quite well. The flowers come in at least seven colours, including a pretty peach shade. This variety does, though, seem to suffer from seed production problems and in some years seed is not available. A bed of this makes a very eye-catching show.

Two other varieties are often seen, 'Jewel' and 'Tom Thumb'. Both tend to be rather more leafy than others with the result that the flowers are hidden amongst the foliage so you end up with lovely green mounds of leaves and no flowers — hardly what you had in mind. One dwarf single colour is regularly seen, 'Empress of India'. This has deep red flowers over rather smaller, dark, sea blue foliage making a lovely combination. In harness with *Helichrysum* 'Limelight' it makes a surprising group.

The one problem with all these varieties occurs in wet seasons and on rich soil. Growth becomes so lush that either the leaf stems grow so long as to overtop the flowers, or the internodes of trailing and climbing types get so long that what flowers you can see are spaced rather sparsely. A well-drained soil that is not too rich is the answer and although they appreciate a liquid feed when planting, and possibly a second soak a week or so later, after that you will not need to water often.

The climbing types can be directed into trees and large shrubs but beware of using too many or the shading effect of the foliage may damage new growth. Rough hedges are good sites for them too. The 'Gleam' type is good on dry banks and in containers of all sorts while the small bushy types are good bedders. The mixtures are best on their own and their dense foliage makes them good weed suppressors in new gardens where the poor soil is likely to suit them too.

Nasturtium seed is very large and easy to handle and so can be space sown outside where the plants are to flower in April. They can be raised in the greenhouse as well and, again, the seed can be space sown in boxes or in cell packs.

There is one other, rather different species, which is grown — *T. peregrinum*, the canary creeper. This is altogether more delicate than the large, coarse hybrids of *T. majus* and has small, pure yellow, slightly frilled flowers and pale foliage. It is perfectly hardy, seeds itself moderately and is ideal for training through shrubs either to add to the display of the shrub or take over later when the shrub itself has finished: 6ft (1.8m) is about the height to expect and almost any shrub that height will make a suitable host. Buddleias and, late in the season, cotoneasters are especially good hosts. The yellow flowers on stems twining round purple buddleia flowers are especially impressive. Seed is best sown where the plants are to flower at any time from March to May.

# U

**Ursinia**

Another in the series of showy annuals in the daisy family, *U. anethoides* is the usual one found. They have the great advantage, unlike a number of other annuals from South Africa, of remaining open throughout the day so you get a really constant display. The foliage is fine, ferny and pale green and the flowers are about 2in (5cm) across. In the 'Mixed' or 'New Hybrids' the flowers will range from rusty orange to lemon yellow, some with a purple disc. They flower from July to September and are happiest in a light but not impoverished soil in good light.

# V

**Vegetables and Herbs**

Now you will think I am really mad when you find me suggesting that you put lettuces and beetroot in the borders with the flowers — and doing it entirely for their looks. But think about it. Garden displays are concerned with colour, form and texture and there really is no reason for one group of plants to be eschewed simply because they are primarily used for something else. At Kew, a large part of the summer bedding display has been given over to vegetables and herbs on at least two occasions and, although some ideas did not work all that well, some were very successful. Many vegetables and herbs can be used, often it is simply a matter of picking the right variety.

*Vegetables*

*Artichokes.* You can make a case for using both globe and Jerusalem types but the former is more valid. Grow it for the grey downy foliage which is very large and well divided and remove any flower spikes to ensure that the energy goes into building up a good crop of leaves. Sow in peat pots, individually, in March and treat as a half-hardy annual, planting out at the end of May. The cardoon has slightly more silvery foliage but is less easy to come by. The Jerusalem artichoke is a very tall plant sometimes reaching 8ft (2.4m) with bright yellow daisies. The variety 'Sunray' is a little more dwarf and rather earlier and more prolific with its flowers. Only useful as a background screen in new gardens.

*Beans.* In spite of a number of varieties of runner beans being suggested as worth growing for their flowers — the red- and white-flowered 'Painted Lady' in particular — I have found that the show of flowers really is not good enough. Again, treat them as a screen that flowers and you will be satisfied. Amongst broad beans dwarf varieties like 'The Sutton' and 'Bonny Lad', from a spring sowing, are very pretty. The flowers on French beans are not impressive but the pods on purple-podded varieties, like the dwarf 'Royal Burgundy' or the climbing 'Purple Podded', make an interesting, long-lasting and attractive display. It helps to make a late sowing amongst the earlier plants. Obviously you will not be removing the pods to eat, and this normally encourages continuous cropping, so the plants tend to slow up in their production.

*Beetroot.* There were once quite a number of varieties of beet grown entirely for their foliage, some varieties had scarlet leaves as much as 3ft (90cm) long but now it is hard to find any at all. The chances are that when it does turn up it will simply be called 'Red Leaved' or some similar purely descriptive name. The leaves are long and narrow and a dark red shade. From a spring sowing pricked out into pots you will get a constant display of foliage all summer long. This is a plant that makes an especially neat edging and also an attractive carpet to set off marigolds or white flowers.

The perpetual spinach beets are good too, especially 'Ruby Chard' and 'Silver Kale', also known as 'Seakale' or 'Swiss Chard'. With both varieties it is the colour of the leaf midribs and veins which are so striking — brilliant red in the former and pure white in the latter. 'Rainbow Chard' is also available with purple, yellow and orange midribbed plants in addition to red and white. The red and white are undoubtedly the most successful and make plants about 12–15in (30–38cm) tall which go very well with lower white or red flowers. Sow in spring like a half-hardy annual but take great care in the pricking out. The seedlings should be moved into pots or cell packs with as much compost on the roots as possible or they can even be sown direct and thinned carefully to one seedling. Disturbance to the roots is liable to discourage them from making a good root system and this in turn leads to bolting soon after planting.

*Brussels sprouts.* Just one variety worth growing here, 'Rubine', which is a lovely, slightly greyish purple shade — not just the buttons but the whole plant. It looks extraordinary in the border surrounded by marigolds or tagetes or you can encourage some canary creeper to run through it. It is most convenient to sow in about March in a cold greenhouse and prick off into trays and then into pots so you have a fairly substantial plant to go out later.

*Cabbage.* You will have guessed that it is the red cabbages that are the best for ornamental work; those especially intended and bred to be ornamental are covered under Brassica. The modern varieties of F1 red cabbage tend to have few loose outer leaves and this is very practical as far as the vegetable grower is concerned. But in the ornamental border a few substantial outer leaves is useful as it adds to the impact of the plant. So pick 'Red Drumhead' rather than 'Ruby Ball'. Try them with low marigolds, *Helichrysum* 'Limelight' or *Pyrethrum* 'Golden Moss'.

*Lettuce*. Two varieties of special interest, the red-leaved and the green-leaved version of the 'Salad Bowl' loose leaf lettuce. Treat them just like any other lettuce you raise in trays and both will make flat, curly mats of foliage in either fresh, clear green or dark bronzy red. Each plant will be 12in (30cm) or more across and is best at the front of the border where the attractive form can easily be seen. They will not bolt, giving a crisp show all summer. The red version is good with red verbena or mimulus while the green makes a good front to the scorching colour of 'Solo' geraniums or 'Laser Purple' salvias. You can try the normal-headed variety, 'Continuity', as well. This too has a reddish tint to the foliage as have a number of continental varieties, including cos types. More of these are slowly becoming available in Britain and would be worth experimenting with although they may not last long without bolting.

*Peas*. Again, like beans, the flowers are not quite good enough but the purple-podded pea, a tall one reaching 5ft (1.5m), is good. Like the purple climbing French bean a late sowing in the same spot will prolong the display.

*Potatoes*. The foliage is not very exciting but at Kew they were grown for their flowers, although the display was rather thin.

*Tomatoes*. A number of varieties have been suggested as ornamentals but of course the problem is that if you leave the ripe fruits on the plants for the sake of their appearance cropping slows down — and you get nothing to go in the salad. The blackbirds eat them too. If you want a tomato to grow in pots on the patio, try 'Totem'.

*Herbs*

*Angelica*. A large biennial making big, bright green leaves heavily divided, with in its second year huge round flower heads, up to 10ft (3m) high. To keep the leaves going and increasing in size and lushness, flower heads can be cut off as they form.

248

*Basil.* The dark purple form 'Dark Opal' is very pretty but needs the right conditions to thrive. Give it a warm, sheltered spot in soil that is not too heavy and make sure it never dries out. Good either with hot or pale colours.

*Borage.* A hardy annual with crisp, bristly foliage and small blue flowers all summer. A good addition to the annual border.

*Parsley.* This is covered earlier in the chapter (see page 213).

**Venidium**

Another of those South African members of the daisy family that are so dramatic in good summers but so disastrous in bad ones. *V. fastuosum* seems to be the only one generally listed and this is a fairly standard half-hardy annual reaching about 2ft (60cm) and carrying long stems of big orange flowers up to 4in (10cm) across. The disc is black and there is a ring of very dark purplish markings around it. An interesting feature of the plant is that the leaves and buds tend to be covered in whitish hairs, like cobwebs. They like a well-drained soil that is fairly fertile, plenty of sun and shelter from the worst of the weather. Give them the usual half-hardy treatment, but prick out into individual 3in (7.5cm) pots rather than into seed boxes. In good years they can be sown outside in late April and make good cut flowers with long stiff stems.

**Verbena**

Once grown only from cuttings for mid-summer bedding schemes, verbenas are now mostly grown from seed, especially by commercial growers. A number have been entered in Fleuroselect trials and the variety 'Tropic' won a Bronze Medal in 1981. Seed-raised verbenas have caught on with home gardeners rather less than with commercial growers for two reasons. First, germination is often less than satisfactory — the crux of the matter is that the compost should not be too wet. If it is reasonably moist when the seed is sown it will not need watering after sowing. The seed should be covered with dryish compost and the tray covered in black polythene and kept in a temperature of around 70°F (21°C). Germination should then be good but there is undoubtedly also a problem with the seed itself, a genetic problem that breeders have not yet solved which reduces germination. If conditions are less than ideal, and home gardeners often find it more difficult to provide perfect conditions, germination may well be lower.

The other problem is mildew and this is usually at its worst on the darker colours, the reds and purples. In bad years this can kill the plants totally by the end of August but precautionary sprays can be very effective. Any fairly good soil suits them as long as it is in no more than half shade and the spreading types are good in troughs and window boxes. The colour range available is now very wide and as well as dark reds and purples includes various pinks, dark and pale blues and whites. There is no yellow but many of the colours come with a sparkling white eye.

There are both upright and spreading types with more and more upright ones appearing. The best of the uprights is the 'Derby' mixture in just four colours — coral pink with a white eye, salmon pink, scarlet with a white eye and rose pink. The mixture is seen most frequently but one or two of the single colours are now to be found. The foliage is deep green and sets off the intense, but not strident, scarlet variety especially well and the rose is one of the most striking of all. 'Tropic' is another more or less erect variety making a plant as wide as high. This was a Fleuroselect medal winner in 1981 and the colour is a unique, deep pink merging to red. A difficult colour to use in the garden but good with purple or deep blue. For a more spreading variety, go for 'Springtime' an older one with flat growth covering quite a wide area and carrying flowers all with white eyes in two pinks, red, pale and dark mauve, purple plus a pure white too. A sparkling variety for trailing out on to paths and patios. 'Blaze' is a good red available separately.

Rather different from these is *V. venosa*. Instead of carrying rounded heads of flowers on divided foliage, this species has deep mauve flowers on rather open plants with narrow leaves. It reaches 12–15in (30–38cm) high and the flowers are set in small, flat, instead of more rounded heads. It is often seen in park and local authority plantings where it is used to create a purple haze over, usually, red geraniums. It is rarely planted in groups and makes but a thin display when treated this way; but in the home garden it is valuable in containers whenever a hint of purple is required. One other variant amongst the verbenas is *V.* 'Aubletia Perfecta'. This grows to 10–12in (25–30cm) and has pale, glossy foliage instead of the more dull and dusky leaves on other varieties. It only comes in one colour, a deep purplish rose pink, but is seemingly immune to mildew and germinates a little more freely.

**Viola (Pansy and Violet)**

There is an enormous range of plants to grow for summer but a much smaller selection to overwinter outside for flowering in spring. So what does the capricious British plant breeder, seedsman and garden centre decide to do? Yes, promote the pansy as a summer-flowering bedding plant, taking one of our best spring flowers and turning it into a summer bedder. You will have realised that this is a trend with which I have little sympathy. Fortunately, a new strain has appeared recently which has encouraged growers again to go in for spring-flowering types.

But first, a little history. It was at the end of the Napoleonic wars that the first interest was shown in improving the humble pansy and creating a more showy garden plant. The wild heartsease is a variable plant both in flower colour and in habit, so keen-eyed head gardeners, encouraged by their aristocratic employers, were able to find a variety of colour forms and bring them into cultivation. At first it was simply the heartsease (*Viola tricolor*) which was grown. This has strong biennial tendencies and grows wild along the edges of meadows and hedges with a particular preference for soil which has been slightly disturbed. It is rather a straggly plant and even today pansies tend to develop in this way if left in the garden too long. Some wild forms are perennial but are still short-lived. Garden forms have been made a little more sturdy by the introduction of blood from *Viola altaica*, a perennial species from Turkestan which arrived in Britain at the beginning of the eighteenth century. This has rather large flowers and helped continue the steady increase in flower size. Colour only changed slowly.

Shortly after the middle of the nineteenth century a different group began to develop. This arose by crossing the more or less annual pansy with another British plant, the mountain pansy, *Viola lutea*. This has a very dwarf, tufted habit and so the resultant plants, those we know as violas, are much more compact in growth, more reliably perennial and the flowers tend to be on longer stalks. It was at this stage that *V. cornuta* from the Pyrenees also became involved.

Named varieties of both pansies and violas, raised from cuttings, are still available but since the 1950s the seed-raised forms of both pansies and violas have dominated increasingly and now there is an enormous number of reliable varieties. In both types there is a range of colours and markings which is unequalled amongst other plants. There is every shade of blue, purple, red, yellow, plus white — with whiskers or without, with blotch (purple, black and now red) or without, tricolours, 'Faces', ruffles, and even black, the nearest you will get to black of any plant. All are best as overwintered plants for spring flowering. For economic reasons there are two ways of treating them — sown and transplanted outside for the less expensive ones while those with more of a tendency to stretch the finances are best treated a little more carefully and sown in a cold frame.

For the more expensive varieties, some of which are F1 hybrids, a method is needed that makes sure no seed is wasted. Of course the outdoor sowing can simply be translated into the protection of frames, cloches or the cold greenhouse soil but the simplest way to treat them is to pot sow in frames. The timing will vary according to how much frame space you have but it would seem convenient to sow after the summer bedding plants have been planted out. But it does depend to some extent on variety and when you want the plants to flower. Seed will germinate fairly quickly in the frame at this time of year and should be sown thinly in 5in (12.5cm) pans or seed trays of peat-based compost. It is important to sow thinly so that when pricking out as much root as possible is transferred with the seedlings. Pansies resent having too much root broken off.

High temperatures cause poor germination so ventilate as much as possible to keep the temperature below 65°F (18°C). It pays to prick out into pots of some sort such as 2½in (6.5cm) square pots and continue growing cool, feeding regularly. The plants can go in their final positions in autumn to flower in spring. If you intend to keep the plants in the frame or cold greenhouse over the winter for planting in spring, sowing can be delayed until August as with the extra protection they will still produce good plants by planting time. On the other hand, by choosing 'Universal' pansies and sowing no later than early July, plants will be in flower by the autumn.

There are only a few acceptable varieties for flowering during the winter. 'Floral Dance' may not have the largest flowers but they are produced in huge quantities. There are seven colours, some with blotches and some without and they stay compact and neat right through the spring too. Recently a lot of publicity has been given to the 'Universal' strain, a group of eleven colours all of which are especially selected for their winter-flowering capacity. The blooms are larger than those of 'Floral Dance' with a greater variety of colours including two where the upper petals are of a different colour to the lower. They flower well right through a mild winter, in average winters there are always flowers on the plants and in hard winters the flowers will still peep through the snow. They are ideal container plants for porches or slightly sheltered corners where they will bloom profusely. Look out for two new ones in this field — 'Spring Fiesta' and 'Diva'.

For spring flowering there is an enormous range of varieties with more appearing all the time. But some of the older varieties are still amongst the best. They are well worth growing in single colours for spring bedding with bulbs and the following are especially floriferous: 'Ullswater' — Wedgwood blue, 'Coronation Gold' — gold/ yellow, 'Snow White' — white with a tiny yellow eye, 'Crimson Queen' — dark red with blotch. As to mixtures for spring flowering, 'Premiere', although not being a good winter-flowering one starts early in the spring, and 'Roggli Swiss Giants' has 13 colours with masks, some with whiskers or contrasting edges, and although they are not the earliest by any means the velvety lustre of their vivid flowers, their unique marking plus the size of their flowers make them an excellent choice. An interesting variety that is relatively new is 'Love Duet'. The background colours vary from lemon yellow, through cream to white with the addition of very pale pink and each has a rosy red mask which also varies slightly in shade. The effect of the slight variation within strict limits and the red mask is very pretty.

In the viola type there have been a number of new types introduced recently which reflects the amount of breeding work being devoted to all the violas. In Europe, America and Japan new varieties are being created and we can look forward to improved winter-flowering capacity in both groups. At present there are a few viola types which stand out. If anything, separate colours are more widely available in viola types. 'Arkwright Ruby' is crimson with a darker mask and is scented, 'Chantreyland' is a stunning apricot, slightly darker towards the centre, 'Campanula Blue' clear sky blue, 'Baby Lucia' has small rounded flowers on long stems with a brilliant yellow eye, 'Red Wing' has dark red upper petals and bright yellow lower petals in similar form to a wild pansy, 'Ruby Queen' has large, pure, velvety red flowers. Then there are the two princes. 'Prince Henry' is an even dark purple and 'Prince John' is pure yellow. Both make compact tufted plants flowering for an exceptionally long period and will still be going strong in the autumn having been planted in the autumn of the previous year. They tend to suffer badly from mildew in dry summers but are otherwise delightful little plants. In mixtures, a very pretty and unusual one stands out, 'Bambini', which comes in a good range of colours most with a contrasting white or yellow face which itself has a dark centre or strong dark whiskers. The variety of faces created makes a very pretty mixture. 'Funny Face' is similar.

For flowering in the summer from a spring sowing a number of varieties have been developed, mainly in pansy types, but in violas too. They are treated as half-hardy annuals but if hardened off well can be planted towards the beginning of May if you have the space to slot them in. High temperatures and drying out during germination can cause problems with spring sown plants so germinate and grow cool and make sure the compost is constantly moist — though not sodden. In the garden try and choose a spot at least partly shaded and with a soil that retains moisture fairly well. The 'Majestic Giants' have very large flowers in seven colours, though all come in groups of shades rather than pure colours. They were bred in America in an area where hot summers are more frequent than in Britain and so do well in most summer seasons. The 'Imperial' strain, also from America, and available so far only in a limited, though expanding, range of single colours is another winner for the summer. 'Imperial Orange Prince' is probably the best of all orange pansies with a dark mahogany whiskered mask, 'Imperial Orange' is a pure apricot, 'Imperial Light Blue' shades from deep blue in the centre to porcelain blue on the upper two petals. Finally there is 'Imperial Pink' which opens a bright purplish pink and fades to a delicate soft pink with darker whiskered face and an orange eye. All are very reliable summer-flowering types. In the viola-flowered varieties, 'Crystal Bowl' has pure flowers with no masks in six colours and the flowers at 2½in (6.5cm) are large for a viola-flowered type. Of course all pansies and violas will produce a reasonable show sown in spring for summer flowering, just as all will do fairly well sown in autumn for spring.

Pansies are amongst the most reliable of spring bedders and their colour range makes them indispensable. Mixed colours look especially well in large containers and raised beds and in large groups in mixed borders. Drifts of single colours together in one bed look superb and this has been done very effectively by the parks department in Southport, and translates very well to the domestic scale. The more subtle colours like 'Love Duet' are especially suited to tubs. Single colours associate well with all bulbs but it is important not to pick bulbs, like Darwin tulips, which are too tall otherwise you end up with a low carpet of flowers, few pansies or violets rise beyond 6in (15cm), then a long length of stem topped by another burst of colour. So pick shorter 'botanical' tulips and try 'Red Riding Hood' with 'Snow White' for a dramatic contrast.

In summer a little protection from the worst of the sun will improve the performance of all varieties though this can be counterbalanced by making sure the soil is in good heart and holds plenty of moisture. In the summer it is the blues and purples that are especially useful, they are colours which are far less common in summer plants than reds and yellows. Try then 'Imperial Light Blue' with marigold 'Solar Sulphur', 'Baby Lucia' makes an interesting show alongside mixed fibrous begonias, or mix them into a 'Novette' impatiens for an unusual display.

## Viscaria

Delightful hardy annuals in the carnation family but resembling them not at all. The viscarias have been shuffled around botanically and have found themselves under silene and lychnis as well as viscaria, where most catalogues put them. They are rather upright, narrow stemmed but stiff plants with grey green foliage and lovely soft flowers. Usual hardy annual treatment is fine, or they can be pot sown and tray grown — but they are not really suited to it. 'Blue Angel' is a slightly variable azure shade on neat compact plants which flower all summer and it has a sister 'Rose Angel', a pale magenta shade. Both have a slightly darker eye and were Highly Commended in the RHS trial of 1983. 'Occulata' comes in a variety of shades in the pink, blue and lilac range and all have a black eye. 'Love' has the largest flowers at about 1½in (4cm). There are mixtures called such things as 'Brilliant Mixture' and

'Treasure Island Mixture' in a lovely range from dark magenta, through various pinks, lilacs and blues to a white with a pink eye. Most reach 12–15in (30–38cm) and make pretty cut flowers as well as good garden plants — use them in big patches.

# X

**Xeranthemum**  Rarely grown, and then usually only for drying, this is also a pretty and easy plant for the border. It grows to about 2ft (60cm) and is rather slender and in need of support. The flowers are single or semi-double, although double varieties in single colours were once available, and are about 1½in (4cm) across. The colour range is limited to white and various lilacs and pinks but they flower for many weeks and are so good for drying that they ought to be much more widely grown. Light, well-drained soil and full sun suits these hardy Mediterraneans.

# Z

**Zea (Maize)**  Ornamental forms of sweet corn are grown either for their foliage or cobs — but it is the foliage forms which are the real winners. 'Japonica' is one you need to look out for but, unfortunately, seed companies are tending to drop it in favour of 'Quadricolour'. 'Japonica' has fresh green foliage with long narrow white stripes — 'Quadricolor' has additional pinkish shades in the leaves which do not improve the effect. They do not flower or fruit so all the energy goes into making masses of long arching leaves giving a dramatically tropical effect. You only need one or two plants — one in a big tub is stunning — otherwise you will overdo it. There are also varieties whose cobs have coloured seeds — red, blue, gold and orange I'm afraid. The idea is that the cobs are dried and varnished for winter decoration.

The plants are not difficult to raise, sow seeds singly in peat pots in a temperature of 70°F (20°C), and plant them out after extra careful hardening off. A good soil is a great help as is plenty of sunshine and shelter from strong wind.

*Figure 7.21*
*Variegated maize is*
*best sown direct*
*into peat pots as the*
*roots resent*
*disturbance*

## Zinnia

The zinnia comes originally from Mexico and more than any of our sun-loving half-hardies really does demand a hot dry summer. In America, especially in the southern States, the situation is of course ideal and this is illustrated by looking at the numbers of zinnias listed by the various seed companies. The one company in Britain whose catalogue also circulates widely in America lists over 30 varieties at the time of writing, whereas none of the others lists more than twelve.

There are three main types grown. The familiar large-flowered, cut flower type comes in a range of brilliant colours and some varieties are also good border plants. Shorter versions are generally also a little smaller in flower size and make good bedders and are ideal in containers as well as pots. Lastly, there is a rather different group of so called 'old fashioned' types with smaller, single or double flowers in large quantities on open, spreading, rather lax plants.

There are two stages at which all groups are vulnerable to the weather: propagation and transplanting. So before looking at the restricted range of varieties that are suitable for Britain, some discussion on how best to treat them is advisable.

Normal half-hardy treatment, pricking out seedlings into trays is the least likely method to be successful. Zinnias not only resent root disturbance but are also especially prone to damping off. The closeness of the seedlings and the difficulty of keeping trays moist but not wet frequently leads to disaster, in spite of precautionary treatment with a copper fungicide This is a method with few merits.

Half-hardy treatment using pots is to be preferred if plants are to be grown as half-hardies. Seeds are sown individually in small peat pots of peat-based compost. If your usual compost contains no sand, perlite or other drainage material then it should be added to help make sure the compost is never too wet. Precautionary treatment against damping off should then be unnecessary. A single seed is sown in each pot and germinated fairly cool, say 60°F (15°C) or even a little less. There is no need to sow before mid-April in most areas. When the seedlings are through and the roots start to penetrate the sides of the peat pots, they are potted on into small pots of the same compost, grown on and hardened off in the usual way.

When it comes to planting out, the site should be in full sun and well-drained and as long as the soil is reasonably fertile no special treatment will be needed. Once planted, water the plants in well once only and then do not water again for some time unless the plants are suffering. Too much water at this stage can easily lead to rotting at soil level. This method is ideal for the cut flower and double-flowered bedding types and as long as they are dead-headed regularly they will produce a reasonable display in most summers.

To some extent the varieties I suggest are based on experience in the especially wet summer of 1985. In such a summer botrytis can be a problem on mature plants of this type. It attacks the flower stems and then spreads further down the plant, as well as sometimes attacking the flowers of the densely double types. Precautionary sprays are advisable here and in practice this means using what's left over from the roses for the zinnias.

All types of zinnias can be treated as hardy annuals and sown where they are to flower in May but this is especially suitable for the 'old fashioned' types which are more likely to be grown in informal groups amongst other annuals or in mixed borders. The usual hardy annual treatment is fine with the special proviso that seed be sown thinly and thinned adequately. The cut flower types are often rather too expensive (20 seeds for the price of a drink) to be treated in this way but they will nevertheless respond well providing cut flowers for display remarkably quickly once they get going. The lack of disturbance together with the relative lack of rainfall during their growing period usually ensures reasonable success. The 'old fashioned' types are especially quick into flower and can be used to fill gaps where something has gone wrong with another variety.

Fluid sowing is also a technique which can be used. This was developed for sowing pre-chitted vegetables outside and can be adapted to other plants. It enables you to gain a week or two while still sowing at the same time. The seed is spread on damp kitchen paper in the bottom of an old plastic ice cream box or margarine tub and kept at a temperature of about 65°F (18°C). When the seed has split it is mixed with wallpaper paste and put in a medium grade polythene bag. The drill is taken out in the normal way, although a little deeper, and watered. The corner is then cut off the bag and the paste/seed mixture squeezed along the row in a fine line. The row is then covered as usual. The great advantage of this system is that you are sowing seed which has already germinated and so can start to grow without the waiting period. This system has not been extensively tried for zinnias but is nevertheless worth a try, especially in colder areas where sowing half-hardies outside has to wait until a little later than elsewhere.

So, which varieties are suitable for the cooler areas which do not enjoy long, hot summers?

In wet seasons none of the tall cut flower types will do well, but the following do the least badly. 'Bouquet' is a chrysanthemum-flowered type with flowers in dark yellow, orange, pink, a beige tinted white and scarlet. Strangely, the scarlet has noticeably larger flowers than the other colours. 'Fruit Bowl', a cactus-flowered type with especially upright flower stems, is also acceptable with a good range of shades. In better seasons the 'Sun' series is excellent. The plants are up to 2ft (60cm) high, the flowers 4in (10cm) across and the range includes red, gold, pink, yellow, carmine, rose and white and these come together in a mixture known as 'Sunshine'. 'Red Sun' and 'Gold Sun' are also to be had separately. Amongst shorter, double-flowered bedding types 'Thumbelina' stands out in wet years, and in general. It grows to just 6in (15cm) and has button-like flowers in varying degrees of doubleness. They will not all be fully double. The colours are pink, orange, yellow, scarlet and white. In wet summers the plants are a little too leafy and there is the occasional plant with longer flower stems and slightly smaller flowers. The 'Dasher Series' is not good in a poor summer but is the best generally, and the red and yellow have been awarded Fleuroselect medals. They reach 12–15in (30–38cm) and the fully double flowers are 3in (7.5cm) across. The number carried will vary from half-a-dozen in bad years to more than twice as many in sunny summers. The colours in the 'Dasher Mixture' are cherry red, orange, pink and white in addition to the scarlet and yellow. Plants in both these groups are worth growing in cold greenhouses or tunnels if you are especially fond of them and your area is not the best favoured.

The 'old fashioned' types are the best for gardens in the cooler zones such as Britain but are very different in style from the taller cut flower varieties. They have more of the hardy annual in their appearance and so are in no way equivalent — except in their cultural requirements. 'Persian Carpet' is the pick in a good or bad season. The plants grow to about 15in (38cm) and are spreading in habit with double and semi-double flowers about 1½–2in (4–5cm) across. The colours are unique. They are all bicolours with the base of each petal being one colour and the tip another. In the 1985 mixture I found eight colour combinations — maroon-purple and orange, mahogany and orange, purple and beige, purple and white, purple and pink, primrose and lilac, mahogany and pale yellow plus one or two all yellows. It flowers quickly from seed so can be treated as a late sown hardy annual. 'Classic', also known as 'Orange Star', has single orange flowers on open, airy plants about 9in (23cm) high and rather more across. Splendid as an edging, in containers and with taller marigolds. Lastly, 'Chippendale', a variety less good in bad seasons, but still better than the big doubles. It reaches 2ft (60cm) and has orange red-centred flowers with a bright yellow edge. Brilliant and dazzling. Most of these 'old fashioned' types will produce long enough stems for the flowers to be cut and their colours make them very useful.

# APPENDIX I
# *Fleuroselect Medal Winners*

The medal winners are chosen from a large number of varieties grown in 22 trial grounds all over Europe from Finland in the north to the south of Italy. About 50–60 varieties are entered each year but the highest number of medals awarded in any one year has been eight. Sometimes only one is given. Since 1973 there have been 39 bronze medals awarded, three silver, but no gold. Not all the varieties are still available.

## Silver Medals

*Geranium* 'Scarlet Diamond'
*Geranium* 'Summer
    Showers'
*Lavatera* 'Silver Cup'

## Bronze Medals

*Alyssum* 'Wonderland'
*Antirrhinum* 'Orange Pixie'
Aster 'Pinocchio'
*Calendula* 'Fiesta Gitana'
*Coreopsis* 'Sunray'
*Dahlia* 'Redskin'
*Dianthus* 'Crimson Charm'
*Dianthus* 'Crimson Knight'
*Dianthus* 'Scarlet Charm'
*Dianthus* 'Scarlet Luminette'
*Dianthus* 'Snowflake'
*Dianthus* 'Telstar Mixed'
*Gazania* 'Mini-Star
    Tangerine'
*Gazania* 'Mini-Star Yellow'
*Geranium* 'Cherry Diamond'
*Geranium* 'Red Elite'
*Geranium* 'Red Express'
*Geranium* 'Sprinter'
*Helichrysum* 'Hot Bikini'
*Impatiens* 'Miss Swiss'
*Lavatera* 'Mont Blanc'

Lavatera 'Tanagra'
Marigold 'Florence'
Marigold 'Honeycomb'
Marigold 'Orange Boy'
Marigold 'Queen Bee'
Marigold 'Showboat Yellow'
Marigold 'Silvia'
Marigold 'Yellow Jacket'
Nicotiana 'Crimson Rock'

Petunia 'Red Picotee'
Rudbeckia 'Goldilocks'
Salvia farinacea 'Victoria'
Verbena 'Tropic'
Zinnia 'Cherry Ruffles'
Zinnia 'Dasher Scarlet'
Zinnia 'Pink Ruffles'
Zinnia 'Yellow Ruffles'
Zinnia 'Pacific Yellow'

## Display Gardens

Displays of Fleuroselect medal winners can be seen at the following gardens:

Harlow Car Gardens, Harrogate.
Sheffield University Botanic Garden.
Royal Horticultural Society Garden, Wisley, Ripley, Surrey.
Sir Thomas and Lady Dixon Park, Belfast.
King's Heath Park, Birmingham.

The annual fleuroselect trials and past medal winners can also be seen at the trial ground run by Unwins Seeds, Histon, Cambridge.

# APPENDIX II
# *All-America Selections*

Started in the 1930s, the All-America selections scheme gives awards to both flower and vegetable varieties. New flowers are assessed at 30 trial grounds all over the United States. There are also over 200 display gardens where winning varieties are on display to the public. Not all varieties are still available.

## Gold Medal Winners

Cosmos 'Sunset'
Linaria 'Fairy Bouquet'
Lobelia 'Rosamund'
Morning Glory 'Scarlet
  O'Hara'
Nasturtium 'Golden Gleam'

Nasturtium 'Scarlet Gleam'
Zinnia 'Peter Pan Pink'
Zinnia 'Peter Pan Plum'
Zinnia 'Scarlet Ruffles'
Zinnia 'Thumbelina'

# Other Winners

*Ageratum* 'Midget Blue'
*Alyssum* 'Rosie O'Day'
*Alyssum* 'Royal Carpet'
Basil 'Ornamental Dark Opal'
Carnation 'Juliet'
Carnation 'Scarlet Luminette'
*Celosia* 'Apricot Brandy'
*Celosia* 'Century Mixed'
*Celosia* 'Fireglow'
*Celosia* 'Golden Triumph'
*Celosia* 'Red Fox'
*Celosia* 'Toreador'
Columbine 'McKana Giant'
Cornflower 'Jubilee Gem'
*Cosmos* 'Dazzler'
*Cosmos* 'Diablo'
*Cynoglossum* 'Firmament'
*Dahlia* 'Redskin'
*Delphinium* 'Connecticut Yankee'
*Dianthus* 'Bravo'
*Dianthus* 'China Doll'
*Dianthus* 'Magic Charms'
*Dianthus* 'Queen of Hearts'
*Dianthus* 'Snowfire'
Foxglove 'Foxy'
*Gazania* 'Mini-Star
*Geranium* 'Rose Diamond'
*Geranium* 'Showgirl'
Gloriosa Daisy 'Double' Tangerine'
*Hibiscus* 'Southern Belle'
Hollyhock 'Marjorette'
Hollyhock 'Silver Puffs'
Hollyhock 'Summer Carnival'
*Hunnemannia* 'Sunlite'
*Impatiens* 'Blitz'
*Kochia* 'Acapulco Silver'
Marigold 'Bolero'
Marigold 'First Lady'
Marigold 'Gold Galore'
Marigold 'Golden Jubilee'
Marigold 'Happy Face'
Marigold 'Janie'
Marigold 'Naughty Marietta'
Marigold 'Orange Jubilee'
Marigold 'Petite Gold'
Marigold 'Petite Harmony'
Marigold 'Petite Orange'
Marigold 'Primrose Lady'
Marigold 'Queen Sophia'
Marigold 'Showboat'
Marigold 'Toreador'
Marigold 'Yellow Galore'
Morning Glory 'Early Call Rose'
Morning Glory 'Pearly Gates'
*Nasturtium* 'Glorious Gleam Mixture'
*Nicotiana* 'Nicki-Red'
*Nierembergia* 'Purple Robe'
Pansy 'Coronation Gold'
Pansy 'Imperial Blue'
Pansy 'Majestic Giant Mixed'
Pansy 'Majestic Giant White with Blotch'
Pansy 'Orange Prince'
Pepper, Ornamental 'Candlelight'
Pepper, Ornamental 'Holiday Cheer'
Pepper, Ornamental 'Holiday Time'
*Petunia* 'Appleblossom'
*Petunia* 'Ballerina'
*Petunia* 'Circus'
*Petunia* 'Comanche'
*Petunia* 'Coral Stain'
*Petunia* 'Glitters'
*Petunia* 'Paleface'
*Petunia* 'Red Picotee'
*Petunia* 'Red Satin'
*Phlox* 'Twinkle'

Snapdragon 'Bright
   Butterflies'
Snapdragon 'Floral Carpet
   Rose'
Snapdragon 'Little Darling'
Snapdragon 'Madame
   Butterfly'
Snapdragon 'Rocket Series'
Snapdragon 'Vanguard'
*Tithonia* 'Torch'
*Verbena* 'Amethyst'
*Verbena* 'Blaze'
*Verbena* 'Trinidad'
*Vinca* 'Polka Dot'
*Zinnia* 'Bonanza'
*Zinnia* 'Border Beauty Rose'
*Zinnia* 'Carved Ivory'
*Zinnia* 'Cherry Buttons'
*Zinnia* 'Cherry Ruffles'

*Zinnia* 'Fantastic Light Pink'
*Zinnia* 'Firecracker'
*Zinnia* 'Gold Sun'
*Zinnia* 'Old Mexico'
*Zinnia* 'Persian Carpet'
*Zinnia* 'Peter Pan Cream'
*Zinnia* 'Peter Pan Flame'
*Zinnia* 'Peter Pan Gold'
*Zinnia* 'Peter Pan Scarlet'
*Zinnia* 'Pink Buttons'
*Zinnia* 'Red Man'
*Zinnia* 'Red Sun'
*Zinnia* 'Rosy Future'
*Zinnia* 'Small World Cherry'
*Zinnia* 'Torch'
*Zinnia* 'Wild Cherry'
*Zinnia* 'Yellow Marvel'
*Zinnia* 'Yellow Ruffles'
*Zinnia* 'Yellow Zenith'

There are a large number of gardens planted with displays of All-America Selections medal winners. For details of your nearest one write to All-America Selections, 628 Executive Drive, Willowbrook, Il 60521.

Fleuroselect medal winners are on display at Michigan and Pennsylvania State Universities.

# APPENDIX III
## *Sources of Supply*

### United Kingdom and Ireland

J. W. Boyce
67 Station Road
Soham Ely
Cambridge CB7 5ED

Chelsea Choice
Folly Farm
Dunmow
Essex CM6 1SG

D. T. Brown & Co
Station Road
Poulton-le-Fylde
Blackpool FY6 7HZ

Chiltern Seeds
Bortree Stile
Ulverston
Cumbria LA12 7PB

Dobies Seeds
Upper Dee Mills
Llangollen
Clywd LL20 8SD

John Chambers
15 Westleigh Road
Barton Seasgrave
Kettering
Northamptonshire NN15 5AJ

Marshalls Seeds
Regal Road
Wisbech
Cambridgeshire PE13 2RF

Mr Fothergill's Seeds
Kentford
Newmarket
Suffolk CB8 7QB

Suffolk Herbs
Sawyers Farm
Little Cornard
Sudbury
Suffolk C010 ONY

Suttons Seeds
Hele Road
Torquay
Devon TQ2 7QJ

Thompson and Morgan Seeds
London Road
Ipswich
Suffolk IP2 0BA

Unwins Seeds
Histon
Cambridge CB4 4LE

## Australia

North Australian Plant
Exports Pty Ltd
247 Tower Road
Suite 12
Montgomery House
Casuarina NT 5792

Southern Cross Seeds
Templestowe Road
Lower Templestowe
Vic 3107

D. Orriell
Seed Exporters Unit
11/10 Golfview Street
Nt Yokine WA 6060

Thompson and Morgan Seeds
Erica Vale Australia Pty Ltd
PO Box 50
Jannali NSW 2226

## Canada

Albion Seed
PO Box 492
Bolton, Ontario LOP 1A0

Apache Seeds Ltd
10136 — 149 Street
Edmonton, Alberta T5P 1L1

Buckerfield's Ltd
1640 Boundary Road
PO Box 7000
Vancouver, BC V6B 4E1

William Dan Seeds
West Hamboro,
Ontario LOR 4A2

265

McFayden
Box 1800
Brandon, Manitoba R7A 6N4

Seed Centre Ltd
Box 3867, Station 'D'
Edmonton, Alberta T5l 4K1

Ontario Seed Co Ltd
16 King Street South
Waterloo, Ontario N2J 3Z9

Stokes Seeds Ltd
PO Box 10
St Catharine's,
Ontario L2R 6R6

W. H. Perron & Co. Ltd
515 Labelle Blvd
Chomedey, Laval,
PQ H7V 2T3

T. & T. Seeds Ltd
PO Box 1710
Winnipeg,
Manitoba R3C 3P6

Robertson-Pike
PO Box 20,000
Edmonton, Alberta
T5J 3M3

Thompson and Morgan Inc.
132 James Avenue East
Winnipeg,
Manitoba R3B ON8

## New Zealand
Arthur Yates & Co
PO Box 940
Auckland

F. Copper Ltd
PO Box 12–347
Penrose, Auckland 1135

## United States of America
W. Atlee Burpee Co.
Warminster, PA 18974

Park Seed Company
Highway 254 N
Greenwood, SC 29647–0001

The Country Garden
Route 2, Box 455A
Crivitz, WI 54114

Stokes Seeds
1036 Stokes Bldg
Buffalo, NY 14240

Chas. C. Hart Seed Co.
PO Box 9169
Wethersfield, CT 06109–0169

Thompson and Morgan Inc
PO Box, 1308
Jackson, New Jersey 08527

Nichols Garden
1190 N. Pacific Hwy
Albany, OR 14240

Harris Seeds
615 Moreton Farm
3670 Buffalo Road
Rochester, NY 14624

# APPENDIX IV
## *Growing Overseas*

Although this book is written from the perspective of the British Isles it should still be of great value to gardeners in other countries. The variation is in the hardiness of the plants described and the timing of operations. The United States Department of Agriculture has divided Britain and the United States into zones of hardiness and Figure A.1 shows exactly where these zones lie. Australia and New Zealand have not yet been covered but the key will give some idea of the temperatures which the individual zones cover. Britain falls almost entirely in zone 8 while North

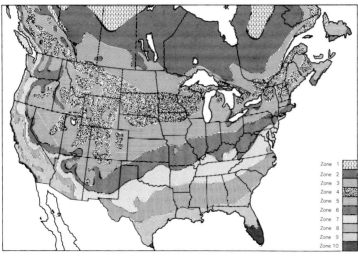

*Figure A.1
Hardiness zone
map of North
America*

Approximate range of average annual minimum
temperatures for each zone

|  | | | | |
|---|---|---|---|---|
| Zone | 1 | below −50 F | (below −45C) |
| Zone | 2 | −50F to | −40F | (−45C to −40C) |
| Zone | 3 | −40F to | −30F | (−40C to −34C) |
| Zone | 4 | −30F to | −20F | (−34C to −29C) |
| Zone | .5 | −20F to | −10F | (−29C to −23C) |
| Zone | 6 | −10F to | 0F | (−23C to −17C) |
| Zone | 7 | 0F to | 10F | (−17C to −12C) |
| Zone | 8 | 10F to | 20F | (−12C to −7C) |
| Zone | 9 | 20F to | 30F | (−7C to −1C) |
| Zone | 10 | 30F to | 40F | (−1C to 5C) |

*Figure A.2
Hardiness zone
map of Great
Britain*

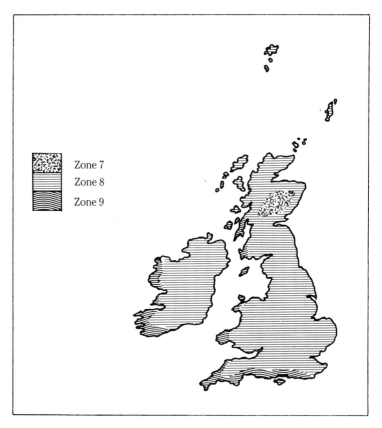

Zone 7
Zone 8
Zone 9

America includes every zone from 1 to 10. But even within zone 8 there is sufficient variation, particularly in the date of the last spring frost and the first autumn frost, to change the method of cultivation in different areas of the zone. It is difficult to be more helpful than to be guided by the hardiness zones, local practice and your own inspiration — and do not hesitate to experiment.

# APPENDIX V
# *Recommended Books*

*Annual Flowers* by Angus Barber (Faber)

*An Encyclopaedia of Annual and Biennial Garden Plants* by Charles O. Booth (Faber)

*Annuals for Garden and Greenhouse* by J. S. Dakers (Collinridge)

*Gardening for Display* by J. R. B. Evison (Collingridge)

*Annuals* by Roy Hay (Bodley Head)

*The Complete Guide to Sweet Peas* by Bernard R. H. Jones (John Gifford)

*The Well-Tempered Garden* by Christopher Lloyd (Penguin)

*Annuals in Colour and Cultivation* by T. C. Mansfield (Collins)

*Weeds and Aliens* by Sir Edward Salisbury (Collins)

*Sweet Peas — Their History, Development and Culture* by Charles W. J. Unwin (Heffer)

Only *The Complete Guide to Sweet Peas, The Well-tempered Garden* and *Weeds and Aliens* are still in print.

# APPENDIX VI
# *Chemical Brand Names*

In this book the names used for garden chemicals are those of the active ingredients. Over the years brand names change and so by asking for a product containing the appropriate active ingredient the right product will be obtained.

However in order to help gardeners in the shorter term, there follows a list of the appropriate brand names as they stand at the time of writing.

Benomyl *ICI Benlate*

Copper carbonate and ammonium carbonate *pbi Cheshunt Compound*

Copper *Murphy Liquid Copper Fungicide*

Derris *pbi Liquid Derris*

Dimethoate *Murphy Systemic Insecticide*

Permethrin *Murphy Tumblebug, ICI Picket, Fisons Whitefly and Caterpillar Killer*

Pirimicarb *ICI Abol–G, ICI Rapid*

Pirimiphos-methyl *ICI Sybol 2*

Propachlor *Murphy Covershield*

Propiconazole *Murphy Tumbleblite*

Thiophanate-methyl *Murphy Systemic Fungicide, May and Baker Fungus Fighter*

# Index

271